Demand Driven Performance

Using Smart Metrics

Debra Smith
Chad Smith

New York Chicago San Francisco
Athens London Madrid
Mexico City Milan New Delhi
Singapore Sydney Toronto

Library of Congress Cataloging-in-Publication Data

Smith, Debra, date.
  Demand driven performance / Debra Smith, Chad Smith.
     pages   cm
  Includes index.
  ISBN 978-0-07-179609-5 (hardback)
  1. Production management—Statistical methods.   2. Industrial management—Statistical
methods.   3. Performance.   4. Costs, Industrial.   5. System theory.   I. Smith, Chad,
date.   II. Title.
  TS155.S578   2014
  658.5′62—dc23

                                                                        2013037670

Demand Driven Performance: Using Smart Metrics

ISBN   978-0-07-179609-5
MHID   0-07-179609-6

The pages within this book were printed on acid-free paper.

Sponsoring Editor Judy Bass	**Proofreader** Amitha Karkera
Acquisitions Coordinator Amy Stonebraker	**Indexer** Robert Swanson
Editorial Supervisor David E. Fogarty	**Production Supervisor** Richard C. Ruzycka
Project Manager Raghavi Khullar, Cenveo® Publisher Services	**Composition** Cenveo Publisher Services
	Art Director, Cover Jeff Weeks
Copy Editor Julie Searls	

*This book is dedicated to the memory of Rudy Harris
and Norm Stevens*

About the Authors

Debra A. Smith, CPA, EMBA, is a cofounder and partner with Constraints Management Group, LLC, a services and technology company specializing in pull-based manufacturing, materials, and project management systems for midrange and large manufacturers. She began her career in public accounting with Touche Ross (now Deloitte Touche), moved to the private sector where she worked as both a controller and a VP of finance for two different publicly traded firms, and then spent seven years as an accounting professor where her research focused on the role of metrics in process improvement. Ms. Smith began working with Dr. Eli Goldratt in 1990. She was elected to the founding Board of Directors of the Theory of Constraints International Certification Organization, serving five years and is certified in all TOC disciplines. Ms. Smith has been active in the Institute of Management Accounting (IMA) and APICS, and a keynote speaker on three continents. She is a coauthor of *The Theory of Constraints and Its Implications for Management Accounting*, the recipient of the 1993 IMA and Price Waterhouse applied research grant, a contributing author for *The Theory of Constraints Handbook*, and the author of *The Measurement Nightmare*. Ms. Smith has been at the forefront of developing and articulating smart metrics.

Chad Smith is a partner with the Demand Driven Institute and is a cofounder and managing partner of Constraints Management Group, LLC. He has been at the forefront of developing and articulating the blueprint for planning and execution in the twenty-first century—Demand Driven MRP (DDMRP). Mr. Smith is the coauthor of *Orlicky's Materials Requirement Planning*, Third Edition, and serves as the Program Director of the Certified Demand Driven Planner (CDDP) Program for the International Supply Chain Education Alliance (ISCEA). He is a frequent consultant to large multinational manufacturing and supply chain companies. Mr. Smith is an internationally recognized expert in the Theory of Constraints and is a contributing author for *The Theory of Constraints Handbook*.

CONTENTS

FOREWORD

You Can Do This—a CEO's Perspective

I can easily identify with your state-of-mind if, after reading this book you still ponder whether the potentials described herein are real. I had these doubts myself at one time. I had previously believed that these types of methodologies would only work in high volume, low mix operations like those in Detroit, or perhaps in other well-defined operations with highly repetitive nature. I had nearly convinced myself that these methodologies could not possibly work at LeTourneau Technologies, Inc. where complexity in operations at our Longview plant was severe (understatement); i.e., highly integrated vertically, from scrap steel to finished product; 600,000 BOM Records; 165,000 Part Numbers; 60,000 Manufacturing Orders annually; 290,000 Manufacturing Operations annually; and 500,000 Drawings. These are WOW numbers by anyone's measure for managing logistics. The logistics challenge is rarely this complex for most companies.

    Why is it so hard for manufacturing executives to capture more of "the gold" from their operations? The short answer is that they often do not see "the gold!" Even when they do see it, they may not believe that it *can* be captured; and, when they do see it, <u>and</u> believe that it *can* be captured, they often do not know how. This is partially explained in that company logistics can be complex, and that our natural behavior does not always lead us to doing the right thing, the right way in such complex situations. In fact, our natural behavior can easily make a complex logistical situation worse; as, there are many examples of this happening in practice. The type of work processes referenced herein for addressing complex logistical situations are not intuitive, and certainly not simple. If they were simple, everyone would be doing them, and doing them right! In this book, the authors will present methods that are proven to work. Once explained, logic will prevail; and, their understanding will become easier.

    Here is further explanation for not capturing "the gold" that relates to how and where management typically focuses. For example, it is completely natural for a manufacturing executive to achieve an effective selling function, as he or she seems to always understand how important this function is to the company's success. Similarly, the same management mind-set normally understands the functional importance of engineering, production, accounting, HR, and IT; and therefore focuses here with expectations. To the contrary, the management of logistics by the same executives is often expected to somehow "happen" without deliberate focus.

This management mind-set frequently looks to the functions of purchasing, production control, and similarly titled functions to inherently perform their supply chain management, and often believes that these functions are producing the best results possible. After all, these functions are likely achieving results comparable to those of the past; and, with an acceptance of past norms being good-enough, this management mind-set fails to see that these seemingly hard working employees almost never achieve the results that may be possible. **Bottom line**: This mind-set in management does not expect, nor demand the achievement of real, breakthrough improvements in their respective supply chains.

Even when and where the achievement of real, breakthrough improvements are expected in a company's supply chain management, it is rare to find leadership that is capable of applying well-thought methods *throughout* the organization, beginning with the top. Pause here for a moment and ponder these basic questions. Where is your CEO on this matter? Is this person visibly involved, leading the charge? We know that this person cares about improving your company results, but have you informed him or her of what is realistically possible? Have you informed this person of how much "gold" there is to be had? And, have you informed this person of how much of the "gold" is not currently being captured?

Doing "the right thing, the right way" in supply chain management clearly requires breaking from convention and tradition. Through managing difficult challenges in an unusually complex case, I know that there are better ways to manage company logistics than what many companies are practicing today. From this experience, I also know with solid confidence that many companies can improve their bottom line by first *seeing* "the gold," and then by *knowing* how to capture "the same gold" with right actions. **Clearly, method improvement should be pursued <u>before</u> spending more Cap-Ex for many companies!** Fortunately in this era, there are proven methodologies and excellent consultant services available that can assist in making your breakthrough improvements happen in practice. This book conveys effective methods that provide many answers in these regards.

A real-world case with LeTourneau Technologies, Inc. (LTI) is only one example where much more of "the gold" was indeed captured in modern times. This improved capture happened through a dramatic shift in management thinking and through much hard work in the path to improvement. This USA manufacturing legend, whose roots date to the 1920s, had developed substantial assets and advantages that became the envy of many heavy equipment manufacturers; i.e., *excellent technologies; strong market participation on six continents; huge production capability; excellent brand name,* and *extraordinary work ethics throughout the organization in essentially all functions and locations.* With this backdrop, one might be prompted to say, "For what more can one ask?" The short answer for LTI became "better logistics management," the kind that makes customers happy and the kind that falls right to the bottom line! Many companies miss "the gold" right here in logistics management.

It is a tribute to the many employees of LTI who helped this manufacturing giant to survive for many decades into its modern setting. Unlike many other USA manufacturing companies who fled offshore, reportedly to survive, the excellent employees at LTI achieved the accomplishments and advantages defined above, right here in the USA. Even with that acknowledgment, these employees admittedly practiced their extraordinary work ethics with a severe handicap; as, LTI had not yet become proficient at logistics management. In fact, we may have rated only "3" on a scale of "0–10," with "10" being what could have been possible through effective logistics management. This failure was portrayed in several negatives, largely summarized as: (1) poor on-time delivery to the customer; and, (2) unacceptable return-on-average capital employed.

This failure to manage logistics occurred in spite of heroic efforts by our employees who had become very good at expediting, pushing hard, working lots of overtime, etc. Remember, this was an organization known for its extraordinary work ethic. They were indeed recognized as the "can do" crowd. In spite of their inherent work behaviors, these same hard-working employees eventually became exhausted at trying their normal intuitive work actions, over and over again, without achieving acceptable results, and without capturing enough of "the gold." It was this exhaustion that led to our ultimate break from convention and tradition in managing logistics. Give us credit at this point in our rich company history for not giving up and for relentlessly believing that there was indeed a better way!

Now, fast-forward to our initial encounter with a highly capable consulting group and their ability to apply a very effective methodology that had not yet become well-known in broad manufacturing circles. We followed their methods much akin to those addressed in this book. Many in the manufacturing world still do not know of these methods; and even where there is awareness of their existence, knowledge and leadership in their execution is lacking. This is not criticism of any person's intelligence; rather, it is an acknowledgment that these implementations are not easy; and again, not intuitive. Among the typical priorities that become routine in many companies, it is often very easy for logistics management to be pushed aside for other "more pressing" matters; and, this makes these types of method implementations even more difficult, with less probability of capturing "the gold."

Fast-forward again, this time to several years later, to the same company operations with the same employees including the same managers, BUT with different methods. For this important LTI case history, there is not enough space herein to address the entire improvement path pursued, including the many associated work actions. Nonetheless, this era is well-documented including its favorable results. While perhaps difficult to believe for anyone who has not yet been engaged in this type improvement work, some of the key results from this era are summarized here: *dramatic improvement in on-time delivery with routine meeting of customer expectations, contrasted from our previous times of rarely meeting customers' delivery expectations; multiple volume growth in new large equipment production, from the same facility that previously was believed to be operating at capacity with lesser volumes; multiples of volume growth in the complex after-market parts business segment that contributed substantially to company profitability; hard-working employees moved from working in frustration to working with more defined purpose and achieving favorable results; return-on average-age capital-employed improved by multiples; and, for the company ownership, **it eventually became possible to sell the company for approximately 25× their net investment after 17 years of ownership.*** These are the facts. Admittedly, as CEO for 13 years, there were many concurrent improvements that contributed to these very favorable results; however, it was clearly the improved management of company logistics that allowed most other improvements to happen. It was indeed better logistics management that was the *theme* improvement that permitted essentially all other improvements. In essence, the benefits of increased volume would <u>not</u> have happened without better management of flow. I believe this strongly with much logic and reason. These same operations had nearly always stuttered with volume surges in the past, but not this time!

Keeping with candor, I acknowledge a failure within our company in the same era by local management at another company location which also had substantial manufacturing capability and many hard-working employees. This local management chose to not use the same methods described herein to pursue their logistics management improvement, and their ensuing results at this second company location became negative. While this was indeed negative, it further proved the authenticity of the improvements achieved simultaneously at the headquarters location. The comparison of these two sets of results within the same company and within the same market era, those *with* effective logistics management <u>and</u> those *without,* proved again that the methods described herein work in practice; and that vice versa, operating results are not as good when not so practiced.

Later, there was another event in the company that further proved the convincing case for the methodologies presented in this book. Shortly after the company sale referenced above, the new owner chose to partially abandon the methods that had been put into practice so successfully at the company headquarters location. This departure from these good methods in one large part of the operations was not mal-intended, but almost certainly a case of not understanding the power of the methods addressed in this book. Almost immediately, the previous excellent results began deteriorating. In essence, the same operations, with their same physical capacities, their same hard working employees, etc., soon slipped into poor on-time deliveries in spite of *lesser volume.* Within the same physical location but in a separate part of these large operations, the business unit managers chose to continue the good methods that had been implemented earlier. The favorable results in this business unit continued. This proved once again that these methods work when properly applied; or vice versa, it proved that favorable results are difficult to achieve in complex logistics management when effective methods are <u>not</u> applied. It admittedly requires "champions" of the improvement cause.

In spite of these very convincing experiences, many companies may continue to operate in their daily ruts, by "doing what they've always done, and getting what they always got." Others, however, may choose to seek real, breakthrough improvements with consistency, and ask, "Why not?" It is indeed a choice, and a very important choice for most. I personally believe that life is too short to not try for breakthrough improvements; and, for those who desire to make such a difference in their respective companies, I recommend the following path of action:

1. Educate yourself by studying the basic literature available on these methodologies.
2. Read and comprehend the contents of *this* book, *after* gaining your basic proficiency above.
3. Become a catalyst to improve your operations with undying propensity to enlist internal support.
4. Seek external help with experience in these methodologies that can expedite your improvements.
5. Engage and convert your top management, by starting with any executive, and not stopping until the highest-level executive leads the implementation with a "stay-the-course" mind-set.
6. Finally, never ever give-up; and, always "go hard." Your company *can* do this.

In closing, as I witnessed our employees succeed in their application of the methods addressed herein, I became a believer. I became well-read on the subject in the process. Admittedly, it took me longer than others, as I had to see to believe. Where have you heard this before? **I can attest that the methods described in this book worked in one of the most complex manufacturing operations that you can imagine with very effective results.** I wish that you could speak with the people to whom this book is dedicated as I personally learned much from Rudy M. Harris who saw our work into fruition. Eventually, I moved beyond being only a believer, to becoming truly passionate about improving company results through applying these methods. Trust me, "this stuff works." I know, from experience!

Dan Eckermann

About Dan Eckermann

Dan Eckermann is a native Texan who spent nearly his entire career working in energy- and mining-related manufacturing. He served on the National Petroleum Council, an advisory body to the Secretary of Energy and is still a passionate "student" of energy. Dan worked for LeTourneau Technologies, Inc. (LTI) for a quarter century, as an operations manager within a business unit, as a business unit head, and then spent his final 13 years as company head (President & CEO) until retirement. During his tenure as President and CEO, LTI grew nearly tenfold.

PREFACE

Einstein once said, "The significant problems we face cannot be solved at the same level of thinking as when we created them." When it comes to the metrics that most of industry uses, the problem that we face is insidious. It permeates nearly everything we do as manufacturing and supply chain entities—it has everything to do with the way we think and behave as organizations. Thus we cannot simply jump into the "right" metrics without first addressing a much deeper issue; the reason why we will fail to see what the "right" metrics are.

Much of industry has lost its way in this more complex and volatile world. Read this book and you will discover when and how the way was lost. Furthermore, you will find a different way to think, design, operate, and measure a system in a more complex and volatile world; a way that is both simpler and smarter.

Metrics tell us how we are doing based on what we want to achieve. Yet it seems most companies struggle to define what they really want to achieve. In the for-profit world it should be relatively simple—the maximization of shareholder equity. The insidious problem referred to above is the route most organizations assume is the way to get there—a route that is totally and unequivocally wrong. The fact that most industries believe that way is right does not make it any less wrong.

Proving the bold statement made above is the journey this book is designed to take the reader on. We believe the book is written in a prerequisite order necessary to fully comprehend the definition and use of what we call *smart metrics*. This book really has three sections. The first section, comprised of Chapters 1 through 4, is designed to first state the problem and then set a framework for managing an organization in the much more complex and volatile world of today. This framework, called the *smarter way*, is comprised of three sequential components. The first component (Chapter 2) addresses how the organization needs to think from a systems perspective. The second component (Chapter 3) describes how manufacturing and supply chain entities must design their operational systems to deal with a more complex and volatile world using the new demand driven operational concepts. The third component (Chapter 4) introduces smart metrics, to sustain and improve these new demand driven systems.

The next section of the book (Chapters 5 through 9) delves into the context and history to understand how convention has resulted in a current suite of metrics that often lead companies in the opposite direction of where they hope to go. Chapter 5 discusses problem analysis through the use of the scientific method in order to reveal a root cause, a deep truth, which has made a very big mess of our metrics. Chapter 6 explores in-depth the implications and assumptions of this root cause. Chapter 7 then takes the reader through how this root cause permeated and persists in our financial and information control systems. Chapter 8 explores the history of financial decision-making in major industries. The breakthroughs and innovations from these past giants reveal key lessons on how to define and find what is really relevant for decision-making. All of the management methodology was created to overcome a dramatic change in technology, markets, and/or consumers; a situation very similar to the dramatic change today that again requires a major innovation in management and metrics. Finally, Chapter 9 concludes with a case study on the Boeing Dreamliner, an illustration of just how big a mess the wrong metrics and assumptions have made of things.

The final section of the book (Chapters 10 through 12) offers a new direction to allow companies to design and operate the new demand driven systems required to remain competitive in today's more complex, integrated, and volatile world. Chapter 10 describes the type of systems modern managers work in today; something called a *complex adaptive system.* These systems require a different way of thinking and a new set of rules than the older, simpler linear systems of yesterday. Chapter 11 introduces a new suite of objectives and metrics in order to keep a complex system balanced—something called *coherence.* Finally, Chapter 12 summarizes the journey and the challenges ahead.

About This Collaboration

In 1997 the authors founded a company called Constraints Management Group, LLC (CMG). CMG was a services company that specialized in implementing and applying the Theory of Constraints in manufacturing companies. CMG encountered immediate success in securing contracts for operational improvement with several large manufacturing companies. A unique combination of finance and metrics and operations expertise combined with thought process tools to help companies see through their chaos prompted these companies to take a chance on this fledgling company.

CMG spent the next six years experiencing success and growth by delivering real, tangible and eye-popping bottom line results through implementing Theory of Constraints in mid-range and large manufacturers. Perhaps the biggest obstacle to bringing these solutions to more companies became the constant obstacle of a client's information system to adequately support the necessary solution design. This lack of information technology elongated implementation timelines as custom solutions often had to be developed in conjunction with a client's IT departments. Attempts at relationships with software providers often resulted in an insufficient solution and a wrestling match for control of the development direction—our vision of how to provide solutions and value simply did not match.

In 2003 CMG entered the software business with plans to build two enterprise software applications. One software application would be used to schedule resources and execute the shop floor while the other overcame the deficiencies of formal planning systems better synchronizing demand signal and supply order generation and execution through even the most complex supply chain scenarios—the types of clients that CMG had demanded nothing short of these objectives. They would also need to provide a performance measurement system based on smart metrics that drive system efficiency and flow.

This pushed the team at CMG to painstakingly articulate the components of the rules behind what would become Demand Driven MRP (DDMRP) and Smart Metrics. DDMRP would change the face of formal planning while Smart Metrics promises to keep those changes sustainable and scalable in a much more volatile and complex world. This articulation and collaboration occurred over years in some of the most diverse and complex industrial environments including working with world leaders in aerospace, mining, oil and gas, food processing and fast moving consumer goods.

The results speak for themselves, read on and discover them or go to the CMG website at: www.thoughtwarepeople.com to learn more.

This is the second book in the Demand Driven series by McGraw-Hill. The first book in the series is the third edition of *Orlicky's Material Requirements Planning* by Carol Ptak and Chad Smith (McGraw-Hill, 2011) in which a blueprint for the future of formal planning, something called *demand driven MRP*, was unveiled. The combination of these two books now gives companies a much more complete picture in order to define, implement, and sustain demand driven concepts.

Debra Smith
Chad Smith

ACKNOWLEDGMENTS

The authors would like to acknowledge a number of special people that they have had the opportunity to work with over the 15 years in order to make this book a reality. First, the late Dr. Eli Goldratt provided a spark that lit a very big fire in both of them. Special executives like Larry VonDeylen, Bruce Janowsky, Dan Eckermann, Dave Woods, David Rex, and Andy McCartney cared deeply enough about their organizations and people to put their careers at risk to drive change in those organizations. Strong operational leaders like Rudy Harris, Norm Stevens, Kris York, Tony Bartellson, Craig Jolly, Joe Marcinski, Joe Moresco, David Mautner, Brad Valentine, Wally Browning, Jacky Williamson, Bill Simon, Pete Johnson, David Blazek, Jeff Stein, Gene Spinks, Jim Stoelk, and Mary Scarcello drove the demand driven tactics to the execution level and then ran interference against the corporate nonsense to keep them in place. The authors would like to thank the exceptional teams at companies like Unilever, Romac Industries, Royal Plastics, Tube Forgings of America, LeTourneau Technologies, Roseburg Forest Products, The Charles Machine Works, Bowden Manufacturing, and Rex Materials Group that moved the needle in a big way for their companies.

The authors would also like to thank the teams at Constraints Management Group LLC, Demand Driven Technologies, and the Demand Driven Institute. Greg Cass, Jeff Herman, Kirk Black, Paddy Rama, George Penzenik, Todd Stafney, Ken Bonnin, Steve Erhlich, and Carol Ptak have worked tirelessly to help define what these concepts are and make them work outside of a textbook.

Chad Smith would like to thank his wife, Sarah, and two daughters Sophia and Lily for putting up with the prolonged absences and locked office door. The support and love of these three people has kept him going. Additionally, Chad would like to thank Carmine Mainiero, Nick Mantenuto, and Malte Schulz for their dedication in bringing demand driven concepts to a global giant in the world of fast moving consumer goods. Chad would also like to acknowledge Erik Bush and his leadership and belief in bringing real and sustainable results to customers. Finally, Chad would like to thank his coauthor Debra Smith for a challenging and rewarding partnership since 1997.

Debra Smith would like to thank her husband Greg for being a partner in every sense of the word for over 30 years and most of all for keeping her laughing. Thanks to her grandchildren Dylan, Sophia, Lily, and Jasmine for making everything new again and putting all things in perspective, and to their mothers Erika and Sarah for bringing them into the world and sharing them. A special thanks to Carol Ptak for being a true friend; 25 years of sharing ideas, feedback, and many bottles of wine, and to Dan Eckermann for living his belief that doing the right thing is goal number one. Finally, Debra would like to thank her coauthor Chad Smith for his incredible ability to know and ask the right question. You keep me on track. You are last here but in my view of the world you're first.

INTRODUCTION: DEEP TRUTH

"This is your last chance. After this, there is no turning back. You take the blue pill—the story ends, you wake up in your bed and believe whatever you want to believe. You take the red pill—you stay in Wonderland and I show you how deep the rabbit hole goes."

> —The character Morpheus in the motion picture
> *The Matrix* (1999, Warner Bros. Pictures)

A Deep Truth lies at the heart of how we perceive reality and how we behave in light of that perception. It is simply what we know. Challenging a Deep Truth is extremely difficult, even perceived as crazy. The Nobel Prize-winning physicist Niels Bohr once said the evidence to replace a Deep Truth must be so compelling, so obvious that people must let go of their attachment to the status quo. In other words, once you see a deeper truth, you simply can't go back. "It is the hallmark of any Deep Truth that its negation is also a Deep Truth."[1]

Today in industry we have a Deep Truth. It permeates all of our operational decision making and behavior. That Deep Truth is the assumption that Return On Investment (ROI) is maximized through and directly corresponds to the minimization of unit cost. Challenging this deep truth can be career limiting. Today, who would stand in front of the CEO and the Board of Directors and say, "We absolutely should not direct our people to minimize unit cost?"

Today's DEEP TRUTH

▼ Unit Cost = ▲ Return on Investment

Everything from curricula approved by academia to the approaches and solutions offered by consulting firms to the major ERP software providers is a part of this Deep Truth. Indeed, entire corporate careers have been built

---

[1]"Discussion with Einstein on Epistemological Problems in Atomic Physics," in *Albert Einstein: Philosopher-Scientist*, P.A. Schilpp (ed.) (1949): p. 240.

around it and devoted to promulgating it. Exposing today's Deep Truth would be threatening to many who are invested heavily in the old and they will act accordingly.

What if today's Deep Truth is totally, completely, unequivocally false?

What if the whole idea of a least unit product cost is simply "bad math"– an inappropriate use of an equation that both economics and even physics would reject?

What if legislation created a reporting requirement that has become the focus of accounting information and replaced, almost by accident, the real definition and rules for relevant information for decision making and product costing?

What if all of our information systems are hard-coded to compile cost reporting and resource area measures from the wrong or misapplied rules and assumptions about how costs and revenue behave?

What if unit cost has become such a Deep Truth that an entire discipline about what defines relevant information has been all but lost?

What if even those who know what relevant costs should be operated inside a system that is not capable of providing relevant information in a relevant time frame to act on?

What if people no longer question taking actions they know will lead to predictable and dire negative consequences that they must deal with later?

The answers to all of these questions will tell us just how deep the rabbit hole is. So which is it, blue pill or red pill?

CHAPTER 1

The Need to Get Smarter

Are We Even Playing by the Right Rules?

Rule [rool] *noun*
1. a principle or regulation governing conduct, action, procedure, arrangement, etc.: *the rules of chess.*[1]

Rules are necessary in any organization. Without rules we have no way to drive and manage behavior. With no way to drive and manage behavior we have corporate chaos. With that said, we need to keep a few things in mind when thinking about rules.

1. **The more rules we have, the greater the chance for conflicting rules.** Modern corporations have rules everywhere and they invest huge amounts of resources in maintaining and enforcing them. In most corporations the number of rules always seems to be expanding.
2. **Conflicting rules are wasteful.** When we have conflicting rules, we have conflicting behaviors. Conflicting behaviors are huge sources of waste as parts of the organization work against each other or themselves. Later we will learn just how pervasive and devastating conflict is throughout organizations.
3. **Many rules are life limited.** Rules are instituted most often based on assumptions about the environment at the time they are made. Rules are often made to accommodate certain limitations. When the assumptions or limitations change, the rules must be re-examined to determine whether they are still appropriate. Souder's Law states that

---

[1]www.dictionary.com

"repetition does not establish validity." Worse yet, the longer the repetition, the more invalid/inappropriate the rule may be.

4. **"Optimizing" inappropriate rules is counterproductive.** Attempts and investment meant to enable or accelerate compliance to inappropriate rules can devastate an organization. If the rule is not only inappropriate but also damaging, then the organization is encouraging its people to do the wrong things faster.

The above points seem relatively obvious, yet many of the mission-critical rules we use to operate manufacturing organizations today . . .

▲ Were conceived more than 40 years ago.
▲ May need serious re-examination.
▲ Often lead to pervasive and widespread conflicts throughout operations.
▲ Often lead to wasteful, even detrimental behavior.

The following sections discuss two mission-critical areas that contain many of these types of rules. These areas are huge players in the manufacturing control and competitiveness equation.

Area 1: Planning and Materials Management

Most midrange and larger manufacturers use a planning method and tool called material requirements planning (MRP). Conceived in the late 1950s, MRP changed the world of manufacturing forever. At that time, computers became more readily available and people realized that computers could make the complex calculations about what to make and buy (given a demand signal input) at a much faster rate than people. A set of rules was created about how those calculations would occur. Those rules remain unchanged as of today in even the most modern enterprise resource planning (ERP) products.[2]

What does it mean? Today's manufacturers are encouraged to calculate demand and generate supply against that demand in the same way they did in the 1960s and 1970s! Is this a problem? Absolutely! Later in this chapter, we will go into detail about what the modern landscape looks like versus 40 years ago. At this point, however, suffice it to say that things have changed significantly.

---

[2]Ptak, Carol, and Chad Smith, *Orlicky's Material Requirements Planning*, 3d ed. New York: McGraw-Hill, 2011, page 40.

ERP planning augmentation with spreadsheets

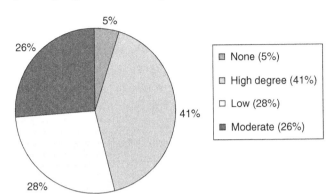

Figure 1.1   Companies Using Spreadsheets to Augment MRP

So, are the rules changing to fit the new circumstances? For the most part, the answer is no.[3] Only now are the relevant professional development organizations and universities beginning to reexamine their core curriculum on this topic. For example, the International Supply Chain Education Alliance (ISCEA) launched the Certified Demand Driven Planner (CDDP) Program in 2012. ERP providers continue to offer the same planning suites but with "advanced" ways to comply with or accommodate the same old rules.

What are the results? Huge amounts of time and energy are spent with little to no return on the investment; advanced forecasting modules don't move the needle on inventory and service levels; training programs teach people to comply with old rules without considering new ones; planners and buyers fundamentally distrust what they see in the system.

Perhaps the biggest indictment of how inappropriate modern planning rules and tools are can be observed in how frequently people work around them. Figure 1.1 shows polling on this subject by the Demand Driven Institute from 2011 to 2012.[4] The poll results are consistent with other surveys by analyst firms such as Aberdeen Group.[5]

This reliance on spreadsheets is often referred to as "Excel hell." Why do planners and buyers do this? Because they know that if they stayed

---

[3] A blueprint for bringing MRP into the twenty-first century was detailed in the third edition of *Orlicky's Material Requirements Planning* (Ptak and Smith, McGraw-Hill, 2011). This new form of MRP is known as demand driven MRP (DDMRP).
[4] Polling sample consisted of over 200 manufacturing organizations.
[5] Aberdeen Group, "Demand Management," Boston: November, 2009.

completely within the rules of the system, approving all recommendations, it would be career limiting. Tomorrow they would undo or reverse half of the things they did today. So what do they do instead? They work around the system. They have their own ways of working with tools that they have crafted and created through their years of experience. These ways of working and tools are unique to each person, with extremely limited ability to be transferred between individuals. This is a different, informal, and highly customized set of rules.

Reliance on these spreadsheets seems to be necessary but comes with risk. In almost all organization there is little to no oversight or auditing of these spreadsheets—the assumption is that the sheets are more sound than the alternative. This assumption, however, can be dangerous. "Close to 90% of spreadsheet documents contain errors, a 2008 analysis of multiple studies suggests. 'Spreadsheets, even after careful development, contain errors in 1% or more of all formula cells,' writes Ray Panko, a professor of IT management at the University of Hawaii and an authority on bad spreadsheet practices. 'In large spreadsheets with thousands of formulas, there will be dozens of undetected errors.'"[6]

As a result of these workarounds, companies and their respective planners and buyers, despite having modern and expensive planning and control systems at their disposal, are committing millions, even billions of dollars of capital every day using:

1. Rules that need serious re-examination,
2. Adhoc workarounds (many of which are error prone), or
3. Some hybrid of both!

Something is horribly wrong with the incredibly expensive planning and materials management systems that we are using. If you are a part of a manufacturing organization, ask the planning and purchasing personnel what would happen if suddenly they had no access to Microsoft Excel and Access products. You probably will not like the answer (unless you also own stock in Microsoft).

Perhaps the more interesting question is, "Why are they allowed to do it?" Why would an organization allow one of its most critical functions to often work outside the system? The answer is simple; survival. To be fair,

[6]Olshan, Jeremy, Wall Street Journal's MarketWatch, April 20th, 2013.

many executives are simply not aware of just how much work is occurring outside the system. Once they become aware, they face a serious dilemma. Let it continue, thus endorsing it by default, or force compliance to a system that your subject matter experts are saying is suspect? The choice is only easy the first time an executive encounters it.

The authors of this book have seen countless examples of executives attempting to end the adhoc systems only to quickly retreat when inventories balloon and service levels tank. They may not understand what's behind the need for the workarounds, but they now know enough to simply get out of the way. So they make the appropriate noises about how the entire company is on the new ERP system and downplay just how much adhoc work is occurring.

Once again, something is horribly wrong with the incredibly expensive planning and materials management systems that we are using.

Area 2: Costing Systems

Today's rules that generate the cost and reporting information to judge performance and make strategic and tactical decisions simply do not reconcile well with what is required to drive ROI in today's environment.

Linear versus Nonlinear

The supply chain systems of today are nonlinear/complex systems. The rules regarding the governance of nonlinear systems are distinct and different from the rules that govern linear systems. The rules for how costs behave between the two are different; many are the opposite. Most business leadership and operational personnel simply do not understand the difference. Thus, conventional costing and reporting information is based on a linear system rule set and the underlying assumption that it should be applied to today's complex manufacturing and supply chains. This assumption is invalid.

Generally Accepted Accounting Principles

Developed in 1934, generally accepted accounting principles (GAAP) is the basis for standard reporting. GAAP is an imposed requirement for the fair presentation of financial statements to external users. GAAP is a forensic snapshot of past performance. This means it is 100% accurate in the past for cost information. It will be 99.9% inaccurate for predicting how costs will specifically behave today and in the future. In today's volatile and complex environment, it is false to assume that the specific circumstances that produced a certain unit cost in the past will be encountered in the exact same way today and in the future. Thus, if companies use GAAP cost information to make planning, execution, and investment decisions today, they are *guaranteed* to use wrong or irrelevant information. When wrong or irrelevant information is used, outcomes will not match expectations. These misalignments in expectations are reflected in the financial statement variances and the difference between planned and actual financial performance.

The failure to recognize supply chains as nonlinear systems combined with the use of GAAP creates and reinforces an assumption about the relationship between cost performance and ROI. We are led to believe that cost improvements everywhere fall to the bottom line. This belief and the pervasive behavior it drives has been hard-coded into all of our information reporting and performance measures. We simply cannot see another way.

The systematized drive to minimize costs leads to just the opposite of its intention; lower service levels, depletion of cash, inflation of inventory, and the squandering of resource capacity and materials. Plant controllers and managers know this; they see it every day. They are constantly placed in conflict between meeting cost performance measures and protecting the other KPIs. They know that if they do nothing but minimize and optimize cost performance, it directly jeopardizes the ROI of the whole system. Ask them what something costs and, before answering, they will ask you why you want to know and what are you going to do with it. In other words, they are trying to determine whether they need to tell the whole story rather than simply give an answer that could lead to trouble. This is one of the biggest indicators that our cost-driven systems and the real world simply do not reconcile.

While GAAP is an imposed requirement to report to external users, a company does not have to impose GAAP on its internal users. GAAP was not built or intended to drive decisions in its manufacturing and supply chain assets—that is the job of a discipline called *management accounting*.

Area 1 + Area 2 = Big Problems

These two areas are troublesome in isolation. In combination, they are devastating. As mentioned previously, the first real MRP systems came online in the 1960s. The MRP systems revolutionized the way companies calculated what to make and what to buy and when to make and buy it. As the use of MRP spread, and the power of computers increased, more and more functionality was added into MRP. In 1972, *closed-loop MRP* integrated capacity scheduling and reconciliation into MRP. In 1980, financials were integrated into MRP, transforming it into manufacturing resources planning (MRP II).

> Manufacturing resource planning (MRP II)—A method for the effective planning of all resources of a manufacturing company. Ideally, it addresses operational planning in units, financial planning in dollars, and has a simulation capability to answer what-if questions. It is made up of a variety of processes, each linked together: business planning, production planning (sales and operations planning), master production scheduling, material requirements planning, capacity requirements planning, and the execution support systems for capacity and material. Output from these systems is integrated with financial reports such as the business plan, purchase commitment report, shipping budget, and inventory projections in dollars. Manufacturing resource planning is a direct outgrowth and extension of closed-loop MRP.[7]

MRP II represents the combined system of hard-coded rules from the two areas described above (planning and costing) and the nail in the coffin for providing relevant costing information to internal managers. Pressure to reduce the cost of middle management stripped out much of the management accounting capabilities of organizations. Today, management accounting has nearly disappeared from the radar screen. This has gone on so long people have come to accept the printout of costing information as real. This problem has been repeatedly pointed out in accounting literature and many other forums for the last two decades.

By 1990, MRP II had evolved into enterprise resources planning (ERP), a bigger, faster, more powerful, and expensive information system. At the core of ERP products today is MRP II and the unchanged and problematic hard-coded rules behind it. This leads to a crucial question: Are we doing the wrong things faster?

---

[7] *APICS Dictionary,* 12th ed. (Blackstone, 2008, page 78).

The Rise of the New Normal

Always Remember: Old Rules + New Circumstances = Expensive Lessons.

Are things really so different today as to drive a need to consider wholesale changes to the mission-critical, hard-coded rules by which our organizations manage themselves? Table 1.1 lists partial circumstances in the industrial landscape that have dramatically changed over the past several decades.

The circumstances in Table 1.1 are key components of the operating environment that manufacturers lived with in 1965 and in 2013. These circumstances should obviously have significant impact on operational (manufacturing and supply chain) design and performance. As these circumstances have changed dramatically, so too must operating models and the rules associated with them.

As an example, consider what Joe Orlicky, the founding father of MRP, wrote about previous inventory and materials management methods in the face of major change—the availability of the computer.

They (previous inventory methods) acted as a crutch and incorporated summary, shortcut, and approximation methods, often based on tenuous or quite unrealistic assumptions, sometimes force-fitting concepts to reality so as to permit the use of a technique.

The most significant results were achieved not by those who chose to improve, refine, and speed up existing procedures, but by those who undertook a fundamental overhaul of their systems.[8]

The irony, of course, is that Orlicky's own argument can now be used against the "new" method that he was introducing to the world in the 1960s and 1970s. In this case, the change is not driven by the removal of a limitation (the speed of computation) but rather the imposition of a set of complex limitations (the New Normal).

Current operating methods and rules *are* based on tenuous or quite unrealistic assumptions. Additionally, they *do* force-fit concepts to a new reality so as to permit the use of an accepted technique.

---

[8]Orlicky, Joseph, *Material Requirements Planning, The New Way of Life in Production and Inventory Management,* New York: McGraw-Hill, 1975, page 4.

Table 1.1   Changing Operating Circumstances from 1965 to 2013

Circumstance	1965	2013
Supply Chain Complexity	**Low.** Supply chains looked like chains—they were more linear. Vertically integrated and domestic supply chains dominated the landscape.	**High.** Supply chains look more like supply webs and are fragmented and extended across the globe.
Product Life Cycles	**Long.** Often measured in years. The rotary phone was untouched for decades.	**Short.** Often measured in months. Don't blink, you might miss the new smartphone launch.
Customer Tolerance Times	**Long.** Often measured in weeks and months.	**Short.** Often measured in days with many situations dictating less than 24-hour turns.
Product Complexity	**Relatively low.** Can you believe that people actually used to work on their own cars?	**High.** Most products now have relatively complex mechanical and electrical systems and microsystems.
Product Customization	**Low.** Few options or custom features available.	**High.** Lots of configuration and customization to a particular customer or customer type.
Product Variety	**Low.** For example, in 1965 Colgate and Crest each made one type of toothpaste.	**High.** In 2012 Colgate made 17 types of toothpaste and Crest made 42!
Long Lead-Time Parts	**Few.** Here the word "long" is in relation to the time the market is willing to wait. By default if customer tolerance times were longer it stands to reason that there were fewer long lead-time parts. More so, supply chains just looked different. Most parts were domestically sourced and thus often much "closer" in time.	**Many.** Today's extended and fragmented supply chains have resulted in not only more purchased items but more purchased items coming from more remote locations.
Forecast Accuracy	**High.** With less variety, longer life cycles and high customer tolerance times, forecast error had relatively little impact—you had time to make corrections.	**Low.** The combined complexity of the above items is making the idea of improving forecast accuracy a quixotic adventure.

(continued)

Table 1.1 Changing Operating Circumstances from 1965 to 2013 (*continued*)

Circumstance	1965	2013
Pressure for Leaner Inventories	**Low.** With less variety and longer cycles the penalties and risks of building inventory positions were minimized.	**High.** At the same time our operations are asked to support a much more complex demand and supply scenario (as defined above) they are asked to do so with less working capital.
Transactional Friction	**High.** Finding suppliers and customers took exhaustive and expensive efforts. Choices were limited. People's first experience with a manufacturer was often through a salesperson sitting in front of them.	**Low.** Information is readily available at the click of a mouse. Choices are almost overwhelming. People's first experience with a manufacturer is often through a screen sitting in front of them.

The Push-and-Promote Problem

The rules embedded in the two areas discussed above (planning and costing) combined to create a mode of operation known as *push-and-promote;* this mode is more supply- and cost-centric than demand-pull centric. Those rules were much more appropriate considering the circumstances under which they were created. Now they represent a real problem, even a threat, to success in the New Normal. The bottom line is that industry has reached, even passed, the point of diminishing returns. Companies that continue to operate using rules rooted in the outdated push-and-promote mode will put more in and get less back.

How can we begin to re-examine what is broken? How can we pull our organizations and their operating models into a more appropriate place?

The Need for Flow

Regardless of the circumstances associated with 1965 or 2013, the recognition of manufacturing as a process is essential to understanding how it should work. Understanding how it should work gives us the capability, in light of current conditions, to define what the rules surrounding it should be. Which rules need to stay? Which need to go? Which need to change? Which need to be added?

George Plossl, a founding father of MRP along with Joe Orlicky and Oliver Wight, once wrote, "manufacturing is a bewildering and distracting variety of products, materials, technology, machines, and people skills obscuring the underlying elegance and simplicity of it as a process. The essence of manufacturing (and supply chain in general) is the flow of materials from suppliers, through plants, through distribution channels to customers, and of information to all parties about what is planned and required, what is happening, what has happened, and what should happen next."[9]

An appreciation of this elegance and simplicity brings us what is known as the first law of manufacturing:

All benefits will be directly related to the speed of flow of information and materials.[10]

A caveat here is that all information and materials must be relevant to the market expectation or requirements. We frequently observe organizations drowning in oceans of data with little relevant information and large stocks of irrelevant materials (too much of the wrong stuff at the wrong time).

"All benefits" is quite an encompassing statement. Let's break it down a bit. All benefits will encompass:

1. **Service.** A system that flows well produces consistent and reliable results. This has implications for meeting customer expectation not only on delivery performance but on quality as well. This is especially true for industries that have shelf-life issues. Do you want to dine at the restaurant that has poor flow or great flow?
2. **Revenue.** When service is consistently high, market share grows or, at a minimum, does not erode.
3. **Inventories.** Raw and pack, work-in-process, and finished goods will be minimized and directly proportional to the amount of time it takes to flow between stages and through the total system. The less time it

---

[9]Plossl, George, *Orlicky's Material Requirements Planning*, 2d ed. New York: McGraw-Hill, 1994, page 4.
[10]Ibid.

takes to flow, the less the total inventory[11] and inventory is captured money, a form of employed capital.

4. **Expenses.** When flow is poor, additional activity and expenses are incurred to close the gaps in flow. Examples would be expedited freight, overtime, rework, cross-shipping, and unplanned partial ships. Most of these activities directly cause cash to leave the organization and are indicative of an inefficient overall system. In many companies these expedite-related expenses are underappreciated and undermeasured—death by a thousand cuts.

5. **Cash.** When flow is maximized, material that a company paid for is converted to cash at a relatively quick and consistent rate. This makes cash flow much easier to manage and predict. Additionally the expedite-related expenses previously mentioned are minimized.

What happens when revenue is maximized and protected, inventory is minimized, and additional and/or unnecessary ancillary expenses are eliminated? Return on investment (ROI) is high. And isn't that really the objective? Every for-*profit* company has a universal primary goal: maximize some form of return on shareholder equity. The best, sustainable way to achieve that goal is to *promote and protect flow*. This is the very definition of an efficient manufacturing system. Conversely, what is one of the fastest ways to compromise ROI and system efficiency? Make decisions and reinforce behaviors that hurt flow.

Once we realize the importance of flow, a few key principles emerge:

1. Time is the ultimate constraint. Time is the most precious resource employed in the manufacturing process. Due to the continual shrinkage of customer tolerance times, this principle is truer today than ever. What we must always keep in mind is that the important time is the time it takes to move through the system. Without this in the front of our mind, we can misuse and distort behavior around time (particularly at the resource level).

2. The system must be well-defined and understood. Clear definition about how materials and information should move will determine whether the system is even capable of maximizing flow.

---

[11]For more understanding of this particular point study Little's Law.

3. Linkages or connections between points in the system must be smooth. Materials and information need to smoothly pass from one point to the other. The greater the friction at these points, the more flow is impeded.

Putting these principles together gives us an elegant point. A company's ability to better manage time and flow **from a systemic perspective** will determine their success in relation to ROI.

It should be noted that the first law *can* work under the push-and-promote mode of operation if and only if the market is willing to support it. As of the writing of this book, that situation can still be found in some of the emerging consumer markets of the world. Ultimately, however, if push-and-promote tactics result in poor flow then that market is ill-suited for those tactics. Remember, rarely can you change the market; frequently you can change your tactics.

Variability—Enemy Number One of Flow

If time is the ultimate constraint and the promotion of flow is the way to best manage it, then we need to understand what is most often impeding our ability to promote flow. What is the primary killer of flow in systems? Simply stated, the answer is variability. The law of variability states, "The more that variability exists in a process, the less productive that process will be."[12]

The concern that the authors have about this definition is that it may not adequately highlight the impact of variability at the system level. The impact of variability must be better understood at the system rather than the discrete process level. The war on variability that has been waged for decades has most often been focused at a discrete process level with little focus or impact to the total system. Variability at a local level in and of itself does not kill system flow. What kills system flow is the accumulation and amplification of variability. Accumulation and amplification happen due to the nature of the system, the manner in which the discrete areas interact (or fail to interact) with each other. Thus we propose a new law:

---

[12]*APICS Dictionary*, 12th ed. (Blackstone, 2008, page 71).

The law of system variability: the more that variability is passed between discrete areas, steps, or processes in a system, the less productive that system will be. The more areas, steps, or processes and connections in the system, the more erosive the effect to system productivity will be.

Figure 1.2 illustrates the law of system variability. The lower half of the graphic depicts a network of connections. It could represent a project network, a bill of material, or even a routing. The point is that it depicts a set of relationships between discrete events, areas, or entities that culminates in some form of completed product, project, or end state. The large squiggly line represents a variability wave that accumulates and amplifies through the system. Delays frequently accumulate, whereas gains rarely accumulate. The graph above the network section shows the impact of the variability wave to system lead time and output. In short, lead time expands while output decays.

Problems associated with variability being passed between discrete areas, steps, or processes are nothing new to the manufacturing and supply chain world. In supply chain, there is something called the *bullwhip effect*.

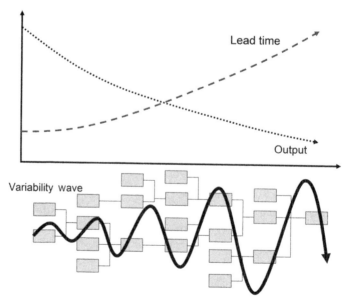

Figure 1.2 Illustrating the Law of System Variability

Demand Signal Changes and Distortions

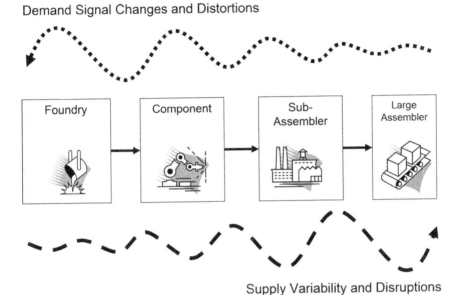

Supply Variability and Disruptions

Figure 1.3 The Bullwhip Effect

Bullwhip effect—An extreme change in the supply position upstream in a supply chain generated by a small change in demand downstream in the supply chain. Inventory can quickly move from being backordered to being excess. This is caused by the serial nature of communicating orders up the chain with the inherent transportation delays of moving product down the chain. The bullwhip effect can be eliminated by synchronizing the supply chain.[13]

The bullwhip is a rather infamous effect in industries with large extended supply chains dominated by major assemblers. Examples include aerospace, automotive, and consumer electronics. Figure 1.3 illustrates the bullwhip effect. Distortions and changes in demand signals move from right to left (customer to supplier) while delays and shortages are passed from left to right (supplier to customer).

A further illustration of system variability occurs in planning systems. It is an effect called *nervousness.*

---

[13]*APICS Dictionary,* 12th ed. (Blackstone, 2008, page 14).

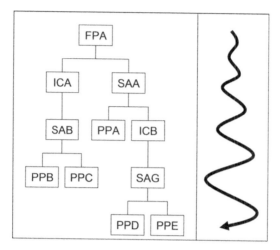

Figure 1.4 System Nervousness

Nervousness—the characteristic in an MRP system when minor changes in higher level (e.g., level 0 or 1) records or the master production schedule cause significant timing or quantity changes in lower level (e.g., level 5 or 6) schedules and orders. Syn: system nervousness.

Nervousness is directly related to the number of dependent connections contained in a company's product structures and how they impact each other. Figure 1.4 illustrates nervousness. As changes occur in quantity and timing requirements at the higher levels in the product structure, those changes automatically impact lower-level quantities and schedules. The squiggly line represents the nervousness running down through the bill of material.

Nervousness is directly related to MRP's biggest power and biggest fault; making everything dependent. Nervousness is often exacerbated with more shared components and materials. Why? There are more dependencies! In fact, nervousness can be so devastating that despite the computing power to easily run real-time or more frequent net requirements calculations,[14] most companies stick to daily or weekly runs. Using the technology to the fullest

---

[14]In MRP, the net requirements for a part or an assembly are derived as a result of applying gross requirements and allocations against inventory on hand, scheduled receipts, and safety stock. Net requirements, lot-sized and offset for lead time, become planned orders. *APICS Dictionary,* 12th ed. (Blackstone, 2008, page 86).

extent possible would produce such a chaotic picture that it would render it useless.

A significant contributing factor to the bullwhip effect in many supply chains is the interaction of the nervousness of the various manufacturing companies that make up the supply chain. Figure 1.5 illustrates nervousness and the bullwhip effect across a supply chain. The leftside of the figure illustrates a supply chain with a large end item producer at the top of it. Each box contains a unique manufacturer with its respective system nervousness on the right. The dotted lines at the bottom of the product structure connect that purchased item to its respective supplier. The right side of the figure depicts the bullwhip effect across the supply chain.

The Types of Variability

If flow is paramount to protect and system variability kills flow, we should explore the nature of system variability encountered by companies. This section is largely taken from the third edition of *Orlicky's Material Requirements Planning* by Carol Ptak and Chad Smith (McGraw-Hill, 2011, pages 16–18).

Variability can be systematically minimized and managed but not eliminated. The biggest challenge in attacking all of the causes for variability and minimizing their individual impact on the system is the investment of time, effort, and money to get there and the return on that investment. The Six Sigma toolset provides an excellent approach to reduce variability at the discrete process level, but even the best master black belt cannot totally eliminate variability (at least not without a magic wand).

From an enterprise perspective, variation has four distinct sources (two internal and two external). These sources come together to create total system variation. The four sources are diagrammed in Fig. 1.6, where the squiggly lines are meant to depict variations that occur within each of those four areas. In Fig. 1.6 the organization is depicted as a pyramid with the management layer in the upper portion of that pyramid. In Chap. 2, this specific shape will make more sense. The output of the organization flows from the bottom of the pyramid.

Demand Variability

Demand variability is an external form of variation. It is characterized by fluctuations and deviations experienced in demand patterns and plans.

18

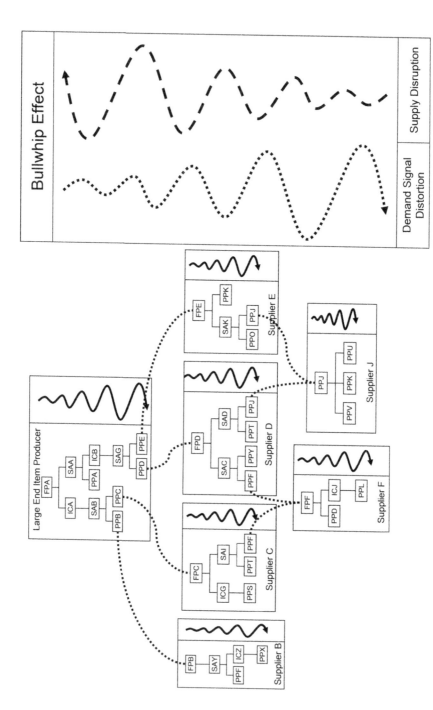

Figure 1.5 System Nervousness and Bullwhip Effect in a Supply Chain

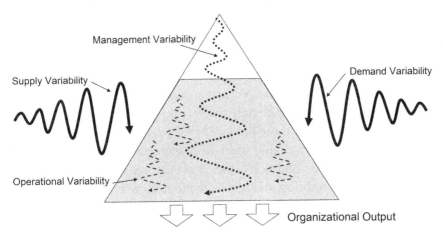

Figure 1.6 The Types of Variation

In Fig. 1.6 it is a squiggly line on the right side of the figure. As discussed previously, in many supply chains, demand variability is driven by the system nervousness of the major players near the top of supply chains. Under the conditions inherent in the New Normal, demand variability and volatility have increased substantially across a much wider range of products.

Supply Variability

Supply variability is an external form of variation. It is a disruption in the supply network or deviation from requested dates and/or promised dates for supply order receipts. It is the reliability (or lack thereof) of the supply network. It only takes one missing part to block the delivery of an end item. In other words, a company can have 99.9% supply reliability and still have unacceptable service levels. In the extreme, a five-cent fastener can block the delivery of a multi-million dollar assembly. In Fig. 1.6, supply variability is the squiggly line entering from the left. Under the conditions of the New Normal, it has increased due to more total suppliers as well as more remote suppliers.

Operational Variability or "Murphy"

Operational variability is an internal form of variation. W. Edwards Deming, the father of Total Quality Management, called this type of variability *common cause of variation.* This is the normal and random variation exhibited by a system in steady state. As mentioned before, perfection at every point of the process is impossible. Even companies embracing the lean approach or utilizing Six Sigma will acknowledge that the desired goal of perfection is impossible. Normal or random operational variability results in a process that may be statistically within calculated control limits but still varying between those limits. The very fact that limits (in plural) are created acknowledges that there is an expectation of variation. In Fig. 1.6 operational variability is depicted as the three small dashed squiggly lines. They are in the operational segment of the pyramid.

Management Variability

Management variability is an internal form of variation. It is variability associated with the human element. Humans make and interpret the rules of a system. Management variability is self-imposed as it is a direct result of decisions made and actions taken within the company. Management variability, even with the best intentions, will frequently take a process out of statistical control (more on this in Chap. 2). This form of variability would be considered by Deming to be a type of special or assignable cause of variability. According to Deming, special cause variability is the first target for improvement. Only when the special cause of variation is addressed can the normal variation of the process be identified. This provides a steady state that is far easier to manage. With that in mind we must tackle this form of variability first.

Getting Smarter—A Basic Blueprint for Change

In the face of increasing system variability, maximizing flow in the New Normal will require change. Defining just what to change and what to change into will require organizations to get smarter. Does that mean that today's organizations are not smart? Organizations have many talented and smart individuals but collectively they are failing to recognize and address the real and fundamental needs for change. How can organizations get collectively smarter?

The blueprint for change is simple; something the authors call "the smarter way." Earlier in this chapter, we wrote that current operating methods and rules *are* based on tenuous or quite unrealistic assumptions and *do* force-fit concepts to a new reality so as to permit the use of an accepted technique. This blueprint for change is pointed directly at addressing management variability; the tenuous and unrealistic assumptions and the damaging habits that create it. We cannot remove the human element, but we can remove the damaging aspect of the human element. In other words, the assignable cause can be addressed given the circumstances of the New Normal. The smarter way has three rather simple steps.

Step 1: Install the Right Thoughtware in the Organization

Are we encouraging and enabling our organizations to think systemically? The authors don't believe so. In fact, decades of combined experience with nearly 1,000 organizations has shown us just the opposite. In most cases, people inside companies are prohibited, discouraged, and/or incapable of thinking about problems and solutions from a systemic point of view (that's a bold statement that Chap. 2 will address). Individuals and the organization can be made capable. To drive meaningful and rapid improvement, problems must be defined and solutions must be developed from a systems and flow based perspective with the New Normal in mind. Organizations have huge obstacles standing in the way of removing the self-imposed variability discussed previously. Chapter 2 will discuss Step 1.

Step 2: Become Demand Driven

The push-and-promote mode of operation must change. In the New Normal, its assumptions and operating methods are square pegs for round holes. Companies will have to find a way to better align their resources and efforts with actual market and customer requirements in the more variable, volatile, and complex environment we have today. Once again, this is not as easy as it sounds as the increasing volatility and variability in the global manufacturing and supply chain landscape make the penalties of guessing wrong and getting it wrong much higher. Chapter 3 will discuss Step 2.

Step 3: Deploy Smart Metrics—Rules for the Smarter Way

At this point the reader may be saying, "Wait a minute! If our organizations are full of the wrong rules, what are the right rules?" An appreciation for what the rules need to be requires Steps 1 and 2. The required changes to sustain competitiveness in the New Normal do require new rules and measures. To embrace and deploy those metrics will require the removal of some very ingrained, even hard-coded assumptions, metrics, and rote behavior. Unless people can think systemically and design operating models to fit the New Normal, these metrics will elude us; the metrics are a function of understanding the fundamental principles of flow and systemic thinking. Chapter 4 will provide an overview of smart metrics. The subsequent chapters will detail the approach, providing a fundamental and elegant set of rules to maximize flow in the New Normal.

CHAPTER 2

Install Thoughtware in the Organization

Variability kills flow and perhaps the most devastating form of variability is self-imposed (more on this later). Even according to Deming, self-imposed variability, management variability, must be tackled first. It is directly related to the assumptions we have and the rules we make and enforce in our system. Our inability to adapt or evolve those assumptions and rules to the circumstances of the New Normal is a recipe for increasing management variability. While the new environment is imposed on the organization, the organization can decide how it operates and responds to it.

This chapter's premise is simple: before an organization should consider making additional huge investments in hardware[1] and software to compensate for the New Normal, it should first consider investing in thoughtware. *Thoughtware* is people's ability to think and communicate systemically. By focusing our collective intuition we will bring to light the problematic assumptions, outdated and conflicting rules, and knee-jerk reactions that not only make our bed but force us to lie on it. Without the correct thoughtware installed, additional investments in hardware and software are often squandered either through their misapplication or due to the fact that they were not really required from a flow perspective in the first place.

Some questions to consider:

1. Are people in your organization formally trained to think systemically?
2. Do they have a common problem-solving language and framework to work within?

---

[1]For our purposes here, hardware includes investments in all capital assets.

3. Do people in your organization understand the connections between departments, resources, and people, or is their intuition limited largely to their silo?

4. Are people given enough visibility to see the relevant connections between departments, resources, and people? Can they convert data to relevant information for flow?

5. Are people discouraged from thinking and offering solutions outside of their area?

6. Can people identify how and where variability accumulates and amplifies to effect total system flow?

Can Our Organizations Think Systemically?

Earlier we made the bold statement that people inside companies are prohibited, discouraged, and/or incapable of thinking about problems and solutions from a systemic point of view. If this is true, organizations can literally be trapped in a cycle of helplessness. How can we make such a claim?

Intuitively almost everyone in an organization knows that the organization is a system, a collection of interdependencies with some sort of common purpose or reason for existence. When it comes to measuring, operating, and problem solving, we tend to divide the organization into subsystems. Those subsystems typically have staffs, budgets, and authority, as well as the expertise to manage within that subsystem. Often instead of really managing a system, we manage a collection of subsystems. The overriding assumption is that if the subsystems are under control then the system will be under control.

Let's take an example. Let's say a company, it could be any industry, has been struggling for quite some time to stay on target to plan. Let's not make it a turn around but the more common failure to meet plan. The CEO calls a meeting in the executive conference room. The entire executive staff is present. In the room are:

▲ Mark, vice president of engineering
▲ Miriam, chief financial officer
▲ Rebecca, vice president of purchasing
▲ Dan, vice president of manufacturing
▲ Johnny, vice president of sales and marketing

The CEO begins the meeting, "We've been given an ultimatum by the board. We either begin to hit our numbers or there are going to be big changes. There are a number of issues we need to address immediately and I need everyone's best effort."

The CEO lists the challenges on a whiteboard:

1. Total sales revenue and market share is down. Not only are we losing customers but we are also experiencing serious price pressure.
2. Service levels are low. We are underperforming in relation to our competitors and it is costing us in market share. We must get service above 95 percent.
3. Product margins are terrible. Our margins have dramatically eroded over the last two years. Pretty soon we are going to be paying people to take our products!
4. Inventory is out of control. We have over 72 days of combined materials, packaging, work-in-process, and finished stocks. The vast majority of it is in materials and packaging. We are turning at fewer than five times per year. All of our competitors claim at least six turns per year. By the way, how can we have this much inventory and still not be able to meet acceptable service levels?
5. Warranty issues have skyrocketed. Last year we had warranty-related costs at over $3,200,000.

How will this meeting end? In the spirit of team problem solving, should the CEO ask everyone's opinion as to why each of these things is occurring and what can be done about it? They will certainly have opinions, explanations, and even an occasional dart to throw at their fellow executives. No, the CEO has been down this road before. The last thing the group needs is an emotionally charged finger-pointing match; this is the time to focus. The CEO must take charge.

"Mark (vice president of engineering), I want your team to do a deep dive into the warranty issues and come to the table with recommendations about a new quality program."

"Miriam (chief financial officer), I need you to look into the product margin issue and make recommendations about what we can immediately do."

"Rebecca (vice president of purchasing), these stock levels are simply unacceptable. I need to see a plan to get these inventories down immediately."

"Dan (vice president of manufacturing), please have your organization come up with a plan to get service above 95 percent."

"Johnny (vice president of sales and marketing), we need to find a way to get our old customers back and get new ones. Please come up with a plan."

What is the chance that each of these people (and their respective staffs) will come up with plans that will work well together? Did they agree on the problem? Can they agree on a common solution or will they all simply "do their own thing" in order to address the issue they have been tasked with?

For example, let's take Rebecca. When she does the analysis she discovers that a large part of the materials and packaging inventories are on slow-moving or obsolete items. Large quantity buys were made (to gain a positive materials cost variance) resulting in bloated positions. Furthermore, it turns out that Engineering made a significant material revision to many of the fast-moving products (in the name of quality). This has further eroded the usage of several materials with high stocks. At their current rate, these materials will not drain out for two and a half years. Packaging inventories is a nightmare. Marketing, without much notification, made significant packaging changes (to support a rebranding effort). Now the old stuff is essentially useless but Finance does not want to acknowledge it, write-off the obsolete stock and take the loss in the current fiscal period when the company is already behind plan.

What can Rebecca do when much of the inventory is unusable or not moving? Ah, but Rebecca does have an option. She *can* dramatically reduce inventory in a very short period of time. She can cancel or defer open supply orders on the materials and packaging that does move quickly. In just a few short weeks Rebecca will make a very big dent in inventory. Will she be praised? Will the bottom line improve?

What will happen? Stocks of critical, fast-moving items will deplete rapidly. Shortages in materials and packaging will block or delay the manufacturing schedule. Service levels will be adversely impacted. Material and packaging expedites will rise dramatically. Flow is reduced to a trickle and Dan (vice president of manufacturing) and Johnny (vice president of sales and marketing) will throw Rebecca under the bus.

This is just Rebecca's story. It is highly likely that the others will have similar stories rooted in implementing an isolated solution meant to treat a particular symptom with little to no consideration of the impacts to other parts of the organization or total system flow.

Remember the ROI equation is proportionate to the speed of flow of materials and information. What kills the speed of flow of materials and information is system variation. Identifying and mitigating system variation takes an understanding and appreciation of how things impact and interact with each other. If an organization's personnel cannot identify and resolve and/or mitigate system variation, then the organization is not collectively thinking about and acting on flow.

Installing thoughtware means creating a framework, context, and opportunity to harness people's intuition in order to improve flow in the system. Many organizations, however, have just the opposite; an operating framework and culture that actively prevents people from generating and considering ideas and taking actions from systems and flow-based perspective.

Thus the primary problem to combat with regard to our organization's ability to think systemically is an organization's and its management's tendency to define problems, construct solutions, and implement plans in isolation.

Referring to Fig. 2.1 (a repeat of Fig. 1.6), let's ask one simple question. Of the four sources of variation, which is under our direct control?

Customer and market behavior (demand variability)? Hardly.

Supplier performance (supply variability)? Only indirectly.

Random events (operational variability)? Please see the definition of *random*.

How we decide to run the business (management variability)? Bingo!

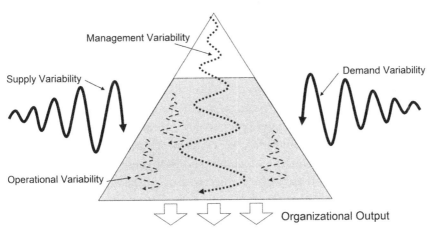

Figure 2.1 The Types of Variation

Once again, this is about management variability: the decisions we make, the framework we create, the culture we foster, and the behavior we reward, all of which are based on assumptions. In fact, if we look deeper into the issue we find that demand, supply, and operational variability are often greatly exacerbated by the assumptions that we have and the corresponding decisions that we make. For example, we can promote products that are fed by temporarily capacity- or material-constrained facilities, which leads to massive service problems. We can give suppliers dramatically different demand signals on a frequent basis or delay paying them on time leading to unplanned material shortages. We can break into the factory schedule on a consistent basis, forcing untold changeovers and confusion leading to quality issues.

Could it be that a significant portion of the variability that we experience as an organization is largely self-inflicted?

Modern Organizations—Conflict Factories

Gasoline (the New Normal), meet Match (management variability). Match, this is Gasoline; enjoy each other. The New Normal pressures us to find ways to dramatically increase organizational efficiency. The contention of this chapter is that almost every company is sitting on a gold mine; it just remains hidden from them. Capturing this hidden but inherent potential will require us to understand the harmful things we do to ourselves, why we do them, and how to stop doing them.

Figure 2.2 is taken from Chap. 14 of *The Theory of Constraints Handbook.*[2] It represents the standard strategic approach to drive ROI, which is the same across all companies we have encountered. No senior manager would argue against any of the components being relevant and necessary for sustained success and healthy ROI; they are the pillars of the ROI equation. The logic is sound: If quality and service go up then sales revenue should go up. If sales revenue goes up and costs go down, then net profit/cash flow should increase. If inventory goes down the invested capital will decrease. Add it all together and ROI goes up.

However, when these tactical objectives are pushed down in the organization, conflicts are created between and within links. The system by

---

[2]Smith, Debra and Herman, Jeff, page 374 (McGraw-Hill, 2010).

Figure 2.2 The Components of Driving Return on Investment

definition is *interdependent,* not independent. The result is often quite messy and chaotic, characterized by finger-pointing, turf wars, and massive amounts of organizational waste. Prove it to yourself by answering the following questions.

▲ Can a drive to increase quality drive up costs and increase cycle time?
▲ Can a drive to decrease cost negatively impact quality and our marketplace?
▲ Can a drive to increase sales erode margins?
▲ Can a drive to increase on-time delivery or shorten our lead time increase cost and inventory and erode quality?
▲ Can programs to decrease inventory starve operations and result in low on-time performance, increased expedite freight, and more overtime?

Most managers make the connections above but they see no way out of the dilemma except to swing from taking actions to maximize one tactical objective to an opposite or competing objective to shore up the objective being jeopardized. This is one of the sources of self-imposed variation. This repeated fire-fighting leads to increases in spending for capital, labor, inventory, and operating expense.[3]

Let's see if we can illustrate just how this happens and just how devastating this is to an organization's performance. The authors must credit Dr. Alan Barnard for first developing the framework of the example shown below. We believe it is one of the best ways to describe how a seemingly

---

[3]Smith, Debra, and Jeff Herman, "Resolving Measurement/Performance Dilemmas," Chapter 14, *Theory of Constraints Handbook,* New York: McGraw-Hill, 2010.

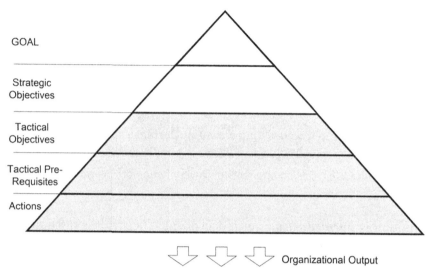

GOAL

Strategic
Objectives

Tactical
Objectives

Tactical Pre-
Requisites

Actions

Organizational Output

Figure 2.3 The Stratified Organizational Pyramid

sound and coherent strategy can consistently create significant and costly operational conflicts.

In Fig. 2.3 we have taken our organizational pyramid and created layers. At the pinnacle of the pyramid is the goal. Just below that is the layer "Strategic Objectives." These are things that are essential to the goal; they are deemed to be necessary conditions. Immediately below that is a layer called "Tactical Objectives." These objectives are commonly broken down by department or area. Below the tactical objectives are "Tactical Prerequisites." These are necessary conditions to the tactical objective—they are commonly associated with primary metrics for the specific tactical area. The final layer is the actions that are taken by department personnel in order to meet these tactical prerequisites.

Using the structure of Fig. 2.3, we will demonstrate how these conflicts occur. Let's consider a situation in which a company is under tremendous pressure; maybe the board is even threatening to step in. The CEO calls a meeting with his or her staff. Perhaps they even travel to an off-site location to all "get on the same page."

The goal is clear. "We need to show a considerably better return for our shareholders over the next year or there will be big changes. What do we need to do?" After lots of thoughtful deliberation, they agree on two

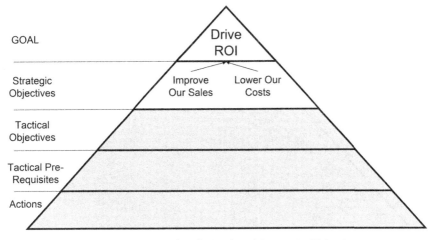

Figure 2.4 Pyramid with Goal and Strategic Objectives

strategic objectives: improve sales and lower costs. What a revelation. Fig. 2.4 has the goal and strategic objectives in the pyramid. The arrows are meant to depict a relationship; what is at the tail of the arrow is deemed to be necessary (not sufficient) to achieve what is at the tip of the arrow. With this in mind, we would say that improving our sales and lowering our costs are necessary to driving better ROI.

Now what should this company do about improving sales and lowering costs? Many ideas were brought to the table. Many were dismissed as conflicting, problematic, or too costly. At the end of a grueling session, they settled on three tactical objectives: improve our offerings to the market, improve our operational performance, and improve our purchasing performance. Figure 2.5 has the pyramid with the tactical objectives in it. In this case, the team is saying that in order to improve sales they must improve their product offerings and improve operational performance. And that to lower costs they must also improve operational performance while also improving purchasing. So far, so good. Things complement each other, even have common components that flow directly up to drive better ROI. This is looking like a great plan.

Now the CEO asks the directors who are closest to these objectives (engineering director, manufacturing director, and materials director) to come up with some key ideas to accomplish them. That evening the three directors sit together and mutually agree on four things that will meet the

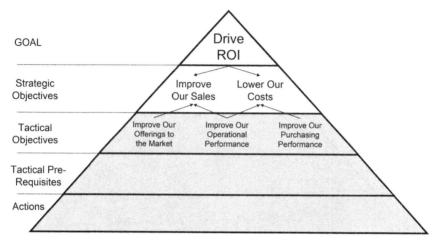

Figure 2.5 Pyramid with Tactical Objectives Added

tactical objectives. In the morning, they present to the group their four ideas: have reliable product quality, improve lead-time performance, ensure material availability, and reduce direct material costs. Figure 2.6 shows these tactical prerequisites in the pyramid.

The directors show how their ideas are not just limited from a departmental perspective but that many are shared in some way, shape, and form between departments. The team is impressed; the directors are congratulated.

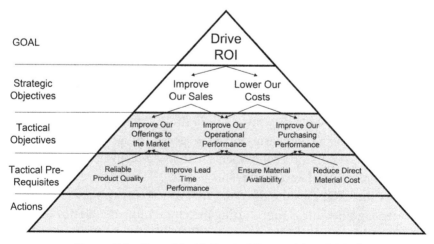

Figure 2.6 Pyramid with Tactical Prerequisites Added

Now the directors need to take this back to their respective teams and make it happen. Meeting adjourned. Let's celebrate!

At first glance, this *is* a reason to celebrate. This plan was constructed cross-functionally, not in silos. That fact alone makes it better than most plans. Unfortunately, at the execution level, actions and interpretations do happen in silos. Watch how a seemingly solid plan results in conflict.

When the purchasing director returns from the executive retreat, a meeting is called with the purchasing staff. "In order to lower our costs and help manufacturing, I have committed to dramatically improve our purchasing performance based on two key criteria: ensuring material availability and the reduction of direct material costs. Please take actions with those criteria as your guide."

When the manufacturing director returns from the executive retreat, a meeting is called with the manufacturing staff. "In order to better support sales and to help lower our costs I have committed to get our lead time down substantially. I know, I know, don't worry, purchasing has agreed to make sure that material is always available. Please take actions to get our manufacturing lead time down."

When the engineering director returns from the executive retreat, a meeting is called with the engineering staff. "In order to better support sales and the costs associated with poor quality, I have committed to better product quality. Please take actions to get much better and more reliable product quality."

Figure 2.7 shows the conflicts within and between departments at the execution level of the organization. Within the purchasing function, there is a conflict about what kind of supplier to source from; the most reliable or the lowest cost. Between Manufacturing and Engineering, there is a conflict about when to release a job to the floor; only after full and complete specifications or in advance of full and complete specifications. Between Engineering and Purchasing, there is a conflict brewing about material substitutions. The lightning symbol signifies a conflict, two opposing actions.

When we explore these conflicts in more depth, a basic structure emerges, a diagram or depiction of conflict. In the Theory of Constraints (TOC), the diagram is called a *conflict cloud*. Having extensive experience in the use of this type of diagram, the authors believe it to be one of the best analytical and communication tools available. Figure 2.8 depicts the conflict cloud structure.

34

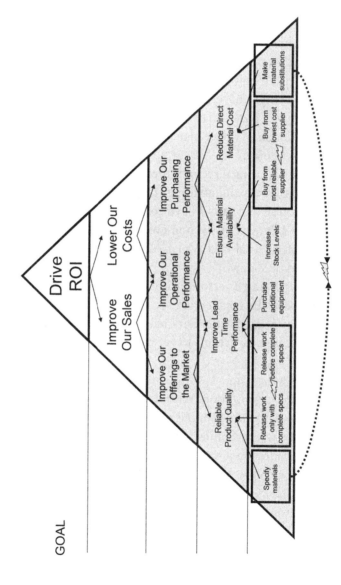

Figure 2.7 Conflict at the Action Level of the Pyramid

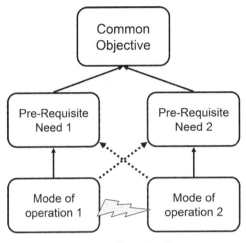

Figure 2.8 Conflict Cloud Structure

Placed at the pinnacle of the diagram is typically an important objective for the system or subsystem, often referred to as the common objective. The common objective is fed by two prerequisite needs. These are things that must be achieved for the objective to occur. The prerequisite needs may not be sufficient in and of themselves, or even together, to make the objective occur; however, if even one of them *does not* occur, the objective will not occur. It is important to note that these prerequisite needs are not in conflict. Remember, they must *both* occur. Where the conflict does take place is between the modes of operation, the actions that these needs drive. The dotted arrow crossing from one mode of operation to the opposite side's prerequisite need is meant to check that a conflict exists. For example, mode of operation 1 compromises or impedes our ability to achieve prerequisite need 2. Mode of operation 2 compromises or impedes our ability to achieve prerequisite need 1.

Let's diagram the three conflicts from the example. Figure 2.9 depicts each of the identified conflicts from the example. The first conflict, labeled "A," is a conflict within a function. In this case purchasing personnel are caught between a rock and a hard spot. On one hand, in order to ensure material availability, they feel pressure to buy from the most reliable supplier; on the other hand, in order to reduce the direct material costs, they feel pressure to buy from the lowest cost supplier. This conflict will not occur on items where the most reliable supplier is also the lowest cost supplier

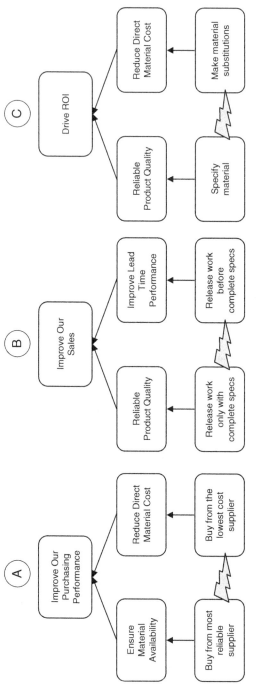

Figure 2.9 The Three Dilemmas from Figure 2.7

(a rare happening). When the conflict is checked, it says that when buying from the most reliable supplier our ability to reduce our direct material cost is impeded; when we buy from the lowest cost supplier it impedes our ability to ensure material availability. When purchasing personnel have formal metrics or explicit orders to do both, the conflict becomes acute.

The next conflict, labeled "B," is between departments; Engineering on the left and Manufacturing on the right. To protect quality, engineering personnel are trying to hold back work release until all specifications are complete, whereas manufacturing personnel are concerned about not having enough time and attempting to release as early as possible. Notice the common objective is not one of the tactical objectives from the pyramid but a strategic objective (i.e., improve our sales). This objective is used because it is the lowest level common objective. Checking this conflict means that when we release work only after full specifications, our ability to improve lead-time performance is compromised, and that when we release work before complete specifications, we jeopardize our ability to have reliable product quality.

The final conflict, labeled "C," is also between departments; Engineering on the left and Purchasing on the right. To protect quality, Engineering is specifying certain materials or specific sources, whereas purchasing personnel are looking for cheaper substitutes. The common objective went all the way to the top of the pyramid, Drive ROI. Checking this conflict for validity means that when Engineering requires a certain material or source, our ability to reduce direct material costs is impeded or compromised, and when Purchasing makes material substitutions the organization's ability to ensure reliable product quality is in jeopardy.

The pyramid example is just a small taste of how rife our organizations are with conflicts. Figure 2.10 is a derivative of a graphic called the *spider web conflict cloud* that appeared in the book *The Measurement Nightmare: How the Theory of Constraints Can Resolve Conflicting Strategies, Policies, and Measures* (CRC Press, 1999, page 25) by Debra Smith. The figure is perhaps more illustrative than the pyramid in depicting just how many conflicts can be ongoing at any one time within and between functions.

Figure 2.10 depicts the shared goal of ROI at the center. The next series of rounded objects all contained in the dotted-line box are the various prerequisite needs to reach that objective. The items within the dotted line are not in conflict; the actions on the outside of the dotted line

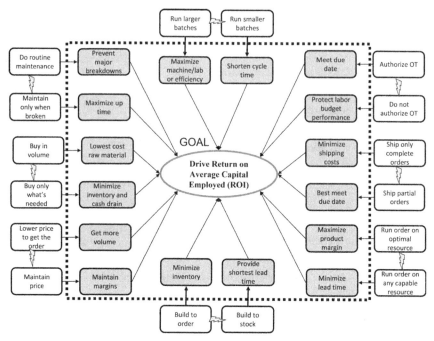

Figure 2.10 A Web of Conflict

frequently are. Those conflicts are illustrated by the lightning symbols. As an exercise, check the conflict clouds by comparing a mode of operation against its opposite prerequisite need. Does that mode of operation block or jeopardize that need?

Conflicts and Oscillation

Conflicts can often produce competitive actions. These actions can waste money and time, and may even poison the working environment. In your home, try having a heater turned on downstairs and an air conditioner turned on upstairs. You will spend a fortune on power and be miserable while doing it. But here is the more important point about these conflicts: conflicting modes of operation typically create oscillations as the power structure and circumstances within and surrounding an organization change. An organization will live on one side of a conflict for a period of time until the negative effects associated with living on that side (many of which are directly related to the sacrifice of the opposite need) force the

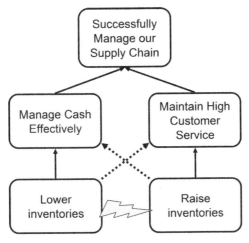

Figure 2.11 The Inventory Conflict

organization to flip to the opposite mode of operation. The self-imposed oscillation is typically delivered via edict, threat, or human sacrifice. Let's use an easy example to illustrate these oscillations.

Figure 2.11 represents a typical conflict with regard to inventory in many manufacturing companies. The common objective is to successfully manage the supply chain. Two very important prerequisite objectives for that are to manage cash effectively and maintain high customer service. In checking the conflict, we ask: Does raising inventories hurt our ability to manage cash effectively? Does lowering inventories frequently jeopardize customer service?

At some point one of the prerequisites becomes more important than the other; let's say customer service. Maybe the CEO has been getting personal (and not-so-nice) phone calls from executives at a large customer. An edict is issued: *Get service up to an acceptable level or else.* This ripples through the organization; people get the point. What's the quickest way to protect availability and fill rates? Raise inventories. We have simply been running too lean!

Figure 2.12 depicts the emphasis on service levels. The prerequisite of high service levels and the mode of operation associated with it immediately come to the forefront while the other prerequisite and its corresponding mode of operation retreat to the background. Finished stock target levels are bumped up, resulting in a surge of manufacturing and purchase orders.

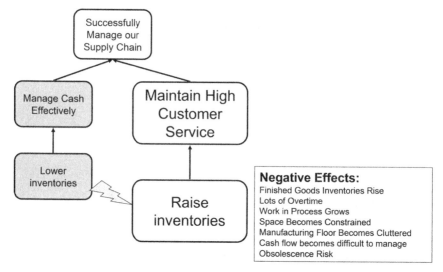

Figure 2.12 The Emphasis on Service

The schedule fills up, overtime is used to combat the flood, large queues of work-in-process accumulate and the floor becomes chaotic and hard to navigate. Finished goods eventually rise and space becomes an issue. Then a routine quality inspection reveals an issue with a critical material; overnight 40 percent of the finished stock is declared contaminated and scrapped.

At this point the CEO has had enough. The organization is cash strapped and under siege from shareholders. People are held accountable; changes are made. The new staff is given clear marching orders: *Clean up this mess, get inventories down, and keep them down.* This ripples through the organization; the survivors get the point and the new blood attacks with vigor. How can we get to a better cash position quickly? Slash inventories, cancel manufacturing and purchase orders, and offer huge price discounts to move the old stuff.

Figure 2.13 depicts the oscillation to the new mode of operation. The prerequisite of cash management comes to the front, while service retreats to the background. It does not take long for material shortages to begin. The organization still has a lot of inventory, but it isn't the right inventory. Manufacturing orders are delayed. Fill rates begin to fall and customer penalties start to kick in. Hot orders to suppliers are launched and expedited in. Crews are held on overtime to wait for inbound material. The manufacturing schedule is constantly shuffled so that we can make and

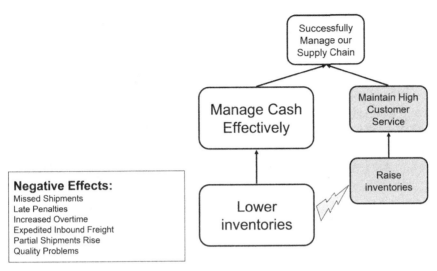

Figure 2.13 The Emphasis on Cash

ship what we can when we can. It is complete chaos on the floor, mistakes are made, and quality becomes a major issue. Customers are furious *and* we are spending a fortune!

Can you predict what will happen next? Will the organization oscillate again? Most likely yes, but that oscillation will probably be led by a new CEO.

A Short Story about Oscillation

The authors were once teaching these conflict definition and resolution techniques to a senior team at a major equipment company, a well-known brand you typically see almost every day. We brought up the point about these oscillations between radically different modes of operation and the effect they have on the organization. We even developed a diagram as a team that mapped out all the effects (called a *current reality tree*). The COO, Dave, got very quiet and basically disengaged for the last hour of the day.

The next morning Dave was the first to arrive at the conference room. He asked if we could speak privately. We sat together in the corner of the room and he brought out two sheets of folded paper from the breast pocket of his sportcoat; printouts of two different emails dated almost 90 days apart. Both represented strongly worded edicts; one to immediately drop inventory and the other to immediately address the rampant service problems.

"I'm the problem," he said. "I thought people weren't doing their jobs but in reality they were; they were doing exactly what I told them to do." Then he asked if he could have some time in front of the group to begin the day.

When the group assembled Dave showed the group the emails. The entire group distinctly remembered those emails. He then apologized to them and told them they deserved better than that. Then he said, "Let's use this opportunity to build a system that prevents this from happening." An hour into the second day of the session, his team was ready to run through a wall for him.

What did this company accomplish by ending the self-destructive oscillations? In less than 12 months, this company dropped inventory by over $35 million, reduced its lead time from 90 days to 14 days, deferred a huge capital investment, and brought nearly $7 million of outsourced business back in-house. All of this was accomplished while growing revenue by 17 percent and without any compromise to service. In fact the service numbers actually rose.

Given these examples, just how much potential is hidden in most organizations?

The Real Definition of Waste

By definition, these oscillations are variability. Worse yet, they are completely self-imposed. We are frequently whipsawing our organizations and its subsystems between two extreme positions. This is costing us more and more as the circumstances of the New Normal cause more frequent oscillations. As these conflicts persist, the organization spins its wheels, even slips backward trading two sets of negative effects and seriously challenging its ability to ever achieve the common objective. This is the absolute epitome of waste: spending lots of time and money to get nowhere. The most fundamental and elegant definition of waste is "unresolved conflict."

Previously, we defined management variability as special or assignable cause variation that frequently takes a system out of steady state. The self-imposed oscillations described above illustrate this. We reiterate Deming's statement that this type of variation must be the first place to attack.

Authors' Note: In the opinions of the authors, the thinking tool set of the Theory of Constraints offers one of the best systemic problem-solving and solution-definition options available.

CHAPTER 3

Becoming Demand Driven

The key to resolving the types of conflicts seen in Chapter 2 lies in the ability to clearly identify and verbalize the underlying assumptions and the rules behind each particular side of those conflicts. Why do people assume that a certain mode of operation is the most expedient way to get to that particular prerequisite need? Is it a belief? Is it a formal metric? Is it an informal metric? Does the assumption, belief, or metric fit the circumstances of the environment or does it indicate a need to re-examine our operating model?

Perhaps the easiest way to resolve or, at the worst, minimize the types of conflicts we have been discussing is to embrace an operating model that is appropriate for the circumstances that we see today. In this chapter, we will discuss moving from the conventional supply- and cost-centric model to a flow and demand pull-centric model. Becoming flow and demand pull-centric is becoming demand driven. Demand driven strategy is about dramatic lead-time compression and the alignment of efforts to respond to market requirements. This includes careful synchronization of planning, scheduling, and execution with actual consumption.[1] Becoming demand driven is essentially forcing a shift from the push-and-promote model of operation to what is known as position and pull.

Push and Promote to Position and Pull

Becoming demand driven involves five steps:

1. Accepting the changes inherent in the New Normal
2. Embracing flow and its implications

---

[1]Ptak and Smith, page 385.

3. Designing an operational model for flow
4. Bringing the demand driven model to the organization
5. Operating and sustaining the demand driven model

Accepting the Changes Inherent in the New Normal

As demonstrated in Chapter 1, the New Normal represents a radically different playing field. Complexity, volatility, and variability (much of it self-imposed) have created extremely complex planning and execution scenarios. The world of manufacturing today looks nothing like it did 30, 40, and 50 years ago when most of the conventional operating methods and rules in the areas of planning and finance were developed. Today those methods continue to be reinforced by software, educational and financial institutions, and a general resistance to big changes. This inertia and stagnation offers an opportunity. Those who are willing to be unconventional but commonsensical can seize it.

Embracing Flow and Its Implications

As explained in Chapter 1, the speed of flow of relevant information and materials directly corresponds to a for-profit company's ultimate goal—maximizing the return on shareholder equity. This connection between flow and return on investment (ROI) has always existed. The circumstances of the New Normal, however, have brought it to the spotlight. Shareholders should be asking fewer questions about costs and instead demand answers about flow.

Designing the Demand Driven Operating Model

Designing a demand driven operating model is about the positioning part of position and pull. To get the positioning right, two things are required:

▲ The identification and placement of decoupling and control points
▲ The consideration of how to protect those decoupling and control points

Placing Decoupling Points

If return is directly related to our ability to protect and promote flow, and variability is the biggest killer of system flow, then we have to design a system that mitigates the effect of variability on flow. The fact that variability exists in discrete processes is not the issue. As discussed before, the accumulation, transference, and amplification of variability really kill flow—what we previously called system variability. With that said, we should amend the first sentence of this paragraph. If return is directly related to our ability to protect and promote flow, and variability is the biggest enemy to system flow, then we have to design a system that mitigates the *transference and amplification* of variability on flow. The inference is that at certain points in our systems we should decouple certain processes; force an element of independence that breaks the variability accumulation chain. This is called a *decoupling point.*

> Decoupling point—The location in the product structure or distribution network where inventory is placed to create independence between processes or entities. Selection of decoupling points is a strategic decision that determines customer lead times and inventory investment.[2]
>
> Decoupling points represent a place to disconnect the events happening on one side from being directly connected to the events happening on the other side. They delineate the boundaries of at least two independently planned and managed horizons. They are most commonly associated with stock positions. As a stock position, they allow demand to accumulate (the stock position drains) but allow for the customers represented by those demand signals to be serviced on demand without incurring the lead-time penalty of the processes in front of the decoupling point. Remember, MRP, our conventional planning tool, struggles with the notion of decoupling points. Its fundamental purpose is to make everything directly dependent.

Decoupling points are not placed everywhere. There are factors to be considered that determine where these points should be strategically placed. The best description of these factors including in-depth and complex examples is in Chapters 4 and 23 in the third edition of *Orlicky's Material Requirements Planning.* In those chapters, Ptak and Smith name

---

[2]*APICS Dictionary,* 12th ed. (Blackstone, 2008, page 34).

Decoupling Point Positioning Factors

Customer Tolerance Time	The amount of time potential customers are willing to wait for the delivery of a good or a service.
Market Potential Lead Time	The lead time that will allow an increase of price or the capture of additional business either through existing or new customer channels.
Demand Variability	The potential for swings and spikes in demand that could overwhelm resources (capacity, stock, cash, etc.).
Supply Variability	The potential for and severity of disruptions in sources of supply and/or specific suppliers. This can also be referred to as supply continuity variability.
Inventory Leverage & Flexibility	The places in the integrated BOM structure (the Matrix BOM) or the distribution network that leave a company with the most available options as well as the best lead-time compression to meet the business needs.
Critical Operation Protection	The minimization of disruption passed to control points, pacesetters or drums.

Figure 3.1 The Six Positioning Factors

what they call six positioning factors. See Fig. 3.1 for a table containing those factors, taken from page 392 of the Ptak and Smith book.

Customer tolerance time (CTT) is the lead time you must maintain to keep business. CTT tells you, at a minimum, how close to the customer you must place a decoupling point to stay competitive. Market potential lead time (MPLT) is the lead time in which you get more business and/or revenue. MPLT tells you how much closer to the customer you should consider placing a decoupling point. The distinction between the two is not trivial. The concept of considering decoupling points within CTT allows a company to explore powerful what-if scenarios with the potential in the market.

Demand and supply variability are the consideration of significant sources of variability, particularly from external sources. Not all end products and materials are equally unreliable and inconsistent.

Inventory leverage and flexibility seeks to consider the aggregate product structure, something called a matrix bill of material. Which shared components matter and which don't, in terms of lead time and working capital compression?

Critical operation protection is the consideration and protection of resources that may be capacity constrained or that easily get knocked out of steady state when variability is passed to them. In some cases, it may require the placement of a formal decoupling point in front of that resource.

Figure 3.2 depicts the conceptual difference between a system with no formal decoupling points and one with formal decoupling points.

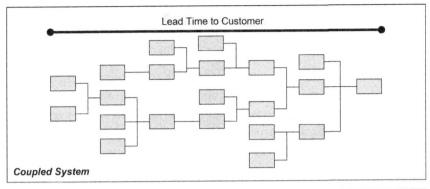

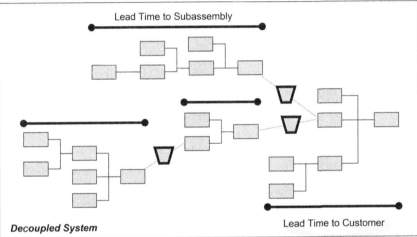

Figure 3.2 Lead-Time Compression with Decoupling Points

The bucket icons in the lower system represent the decoupling points. The dotted lines running through the decoupling points are meant to represent the indirect connections between the two independently planned and managed sides of the decoupling point.

It is important to note that:

1. The lead times are decoupled.
2. Adding the longest path of decoupled lead times still produces a similar lead-time number as the coupled system. The crucial difference is that the lead time that the customer reliably experiences is tremendously shorter, which can be a significant market advantage.

3. This has huge implications for planning. If the planning lead time shrinks, then the forecast error over the planning lead time also shrinks as forecast error rate grows as the length of the planning horizon elongates.

It should be noted that an entire discipline about the placement, protection, and operation of these decoupling points within and across manufacturers and distribution networks has been articulated and proven. This discipline is called demand driven MRP (DDMRP). It is thoroughly documented in the third edition of the Orlicky text cited above.

Placing Control Points

Control points should be thought of as places *to transfer and amplify control* through the system. Control points are often placed between decoupling points with the objective of better controlling the lead-time zones between those decoupling points. A shorter lead time and less variability within a zone leads to less required stock at the decoupling point (a working capital reduction).

Many situations do not allow or require the decoupling point at the end item level (between the customer and the last stage in the manufacturing process). For example, in an engineer-to-order environment, every end item is unique to the customer—an end item cannot be stocked. Certain subassemblies and/or materials could have decoupling points, but not the end item. Another example would be when the customer tolerance time is longer than the lead time from the last established decoupling point to shipping. Figure 3.2 actually shows this situation, as there is still significant activity after the last decoupling points. In this case, a control point (maybe more than one) will be established between those last decoupling points and delivery to the customer.

> Control points—Strategic locations in the logical product structure for a product or family that simplify the planning, scheduling, and control functions. Control points include gating operations, convergent points, divergent points, constraints, and shipping points. Detailed scheduling instructions are planned, implemented, and monitored at these locations.[3]

---

[3] Ibid.

Instead of attempting to control the system through the management of every minute of every resource, meaningful control over a group can be asserted and maintained from relatively few places. An example might be security at an airport. While surveillance is occurring everywhere, active control is asserted at only a few points. From those few points, security across hundreds of flights and tens of thousands of people can be extended with minimal disruption. Imagine the cost, chaos, delays, and risks associated with security being performed for each flight at both the ticket counter and the gate.

Control points don't decouple lead times; they seek to better manage the lead-time horizons in which they are directly involved. Control points are the first areas to be scheduled based on a requested final completion time (either the delivery to a customer or to a decoupling point). The control point schedule then drives all other resource and area schedules within that lead-time horizon, creating a staggering effect for material release and scheduled completions (promise dates). In the Theory of Constraints, control points are called *drums*, because they set the cadence of the system. In Lean, control points are often called *pacesetters*.

Where to place control points? As with decoupling points, there are considerations that should be made when choosing where a control point should be placed.

Consideration 1: Points of Scarce Capacity

Anyone who has read the business novel, *The Goal*, by Dr. Eli Goldratt will probably remember "Herbie," the slowest Boy Scout in the troop. The logic is simple. The total system output is limited to the slowest resource. If demand exceeds or nearly exceeds that resource's ability to produce, then any time lost at that resource is essentially time (dollar generation potential) lost forever. Figure 3.3 represents what a capacity constrained control point candidate would look like when comparing available capacity (y-axis) to required load (the dotted line) across a spectrum of resource centers (A through K). Resource D appears to be an ideal control point candidate. Resource D has the least available capacity to required load ratio. While it is not a bottleneck (a resource whose required load exceeds its available capacity), it is a capacity constrained resource (a resource with limited capacity that, if disrupted, could become a bottleneck). This resource must be carefully and finitely scheduled and that schedule must be maintained.

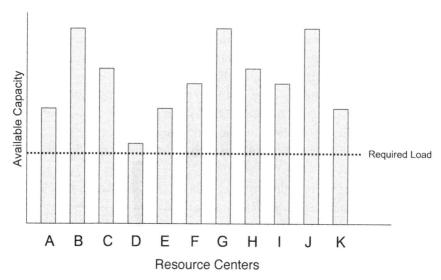

Figure 3.3 Load versus Capacity Analysis

Disruptions to the schedule may cause this resource's limited capacity to be squandered, rendering it a bottleneck.

This particular consideration (capacity) is so obvious and has been made so famous by *The Goal*, it is almost detrimental. Companies should be cautioned: if a capacity constrained resource is not evident or present, it *does not* mean there should be no control point. The following considerations still apply.

Consideration 2: Exit and Entry Points

Establishing a schedule of exit and entry (gates) allows a company to better establish control over the entire process; it allows us to establish limits to live within. These control points obviously include the release or introduction of work and the delivery of finished work to a decoupling point or customer. This consideration also includes the interaction with an outside processor. When work leaves our system, it leaves our effective control. Thus, we should establish control limits (schedules) for that gap in control and visibility.

Consideration 3: Common Points

Common points are points where product structures or manufacturing routings either come together (converge) or deviate (diverge). The power of these types of places should not be underestimated. It has nothing to do with what work content is performed there. It has everything to do with the point's position in the sequence or hierarchy of the part routing or structure. If there is a place that many paths start from or converge to, then *all* of those paths can be controlled to some extent from that one place. They provide tremendous leverage opportunity and are focal points for visibility to and information on the status of flow.

Sometimes areas that fit this criterion can be deemed non-value- or little-value-added operations staffed by minimum wage workers. As an example, let's continue the story from Chapter 2 about Dave's company (the COO who realized he was forcing oscillations through his company). His company had over 1 million square feet of manufacturing space and over 1,500 direct employees on the manufacturing floor. The plant was essentially divided into two distinct sides. On one side was a massive fabrication and machining operation with many diverse types of resources requiring highly skilled labor. More than 700 people were on this side of the plant, with tons of parallel activity. Work could be routed through many alternative routings to achieve the same result. How to assert control in such a flexible environment? Should control points be placed everywhere?

Then someone asked a seemingly stupid question, "What about the paint line?" The objections started immediately. The paint line ran less than two shifts. It was old and dirty. The labor was unskilled—"they just hang things on hooks!" All that was true, but more than 90 percent of the items that went through the front side of the plant were painted *one* color in *one* area. As a control point, the paint line could synchronize the flow of a massive amount of activity. After several hours of "yeah, buts. . . ," and, "what ifs. . . ," the team agreed to try it. The founder of the company, in his eighties at the time, a brilliant engineer with more than 100 patents to his name, simply chuckled when presented with the idea. He said two words, "How elegant."

Consideration 4: Points That Have Notorious Process Instability

When a resource is wildly out of steady state, making it a control point forces the organization to get it under control. These resources are often simply not well understood by the organization. What does the resource really need for an input? How does it really produce the output? What is its clearly defined purpose? Maybe the organization has lost the personnel who understood how to make the resource function properly. This misunderstanding can cause the resource to be misused, leading to instability and variable quality.

This situation occurred while the authors were working with the world's largest diversified freeze dryer, Oregon Freeze Dry. Oregon Freeze Dry had seen tremendous benefit from the use of decoupling points between the plant and their customers' warehouses. Now they wanted to deploy a solution inside their plants to dramatically reduce their planning horizon from the current two weeks down to two days. But where to place the control point? The drying chambers were by far the most heavily capitalized assets and expensive to operate. They were not going to build more drying chambers anytime in the near future. It seemed to make sense to declare the chambers control points and finitely schedule them, pacing the rest of the resources to those schedules. Schedules that would maximize their potential flow.

The problem with that strategy was exposed when the capability of the preceding step in the process was explored. Before you can dry something, you must first freeze it.[4] This freezing process occurs in what is called a "cold room." A cold room is a large insulated room typically held at a temperature well below freezing. The unfrozen food (either a raw or cooked mixture) is placed on racks of trays on a large cart. The cold room has large doors that can quickly open for cart entry and exit. At the time, Oregon Freeze Dry had four cold rooms and had plans for a fifth (a sizable investment in capital and an expensive piece of equipment to operate). If the cold rooms frequently disrupted the chamber schedules with today's two-week planning horizon, what would happen with a two-day horizon?

Why do the cold rooms frequently disrupt the chambers today? The cold rooms frequently have to be shut down due to frost buildup, and food scrapped due to poor quality resulting from inconsistent freeze rates. The cold rooms seemed to be a mystery. Sometimes the quality was fine, and frost was not an issue. Other times, running many of the same products, it was just the opposite.

Further examination led to the discovery that the cold rooms were frequently jam packed with carts. The doors had to stay open longer in order to find a spot to put the cart or to find the right cart to retrieve from

---

[4]Freeze drying is a method to dry various materials while preserving their physical structure. Material is first frozen and then warmed in a vacuum chamber. The water in the material passes from a solid state directly to a gaseous state (sublimation).

the cold room. Obviously, this resulted in temperature instability, which led to some of the quality issues as well as extended freeze durations. But why was the cold room always full?

Further inquiries revealed that the way the company thought about the capacity of the cold room was tied to available space (square footage). If there was space available in the cold room, larger batches were produced and stored. Larger batches meant more carts going in. More carts going in made the door open more frequently. As space became constrained, the door stayed open longer both to deliver and retrieve carts. Furthermore, larger batches of hot products (stroganoff or macaroni and cheese) introduced huge temperature swings in cold rooms, resulting in extended freeze times, poor quality, frost buildup, and eventual cold room defrosting.

This forced a fundamental question to be asked: What is the point of a cold room? The answer was to bring material to a desired and consistent frozen state, ready for drying. How does a cold room do that? It must take heat out of the material at a consistent rate. The chillers in the cold room must be able to support that rate of heat transfer. In other words, it has nothing to do with how much floor space is available in the room but how much available chiller capacity remains to maintain temperature. Now when the cold rooms are loaded with hot products the load on the chiller is greater and less of the square footage is used. Freeze rates and quality will now be consistent. This means when the cold room is freezing broccoli or corn, the room may have more carts in it (more square footage used).

It took about 1 minute for the operational leadership to get it. In a company full of engineers and physicists, this was a big "oops!" The focus had always been on the drying chambers—that's where all the cutting-edge science, intellectual property, and attention was placed. At lunch Larry Von Deylen, the senior vice president of operations said, "I'm embarrassed to say that we have been in the freeze drying business for nearly 30 years and it appears that we knew very little about freezing."

Temporarily establishing the cold rooms as a control point and scheduling them to chiller load produced quick and dramatic results. Freeze times and quality consistency improved dramatically. The company quickly realized that only two cold rooms were needed. They tabled the plans for a fifth cold room and converted the two unneeded rooms into refrigeration capacity for raw materials. Previously they had been leasing refrigeration capacity from an external entity. That spend was recaptured and the raw material was now on-site.

A Decoupling and Control Point Example

In summary, strategically placed decoupling and control points can dramatically compress lead times to meet market requirements and/or opportunities and assert or impose control throughout the system. To illustrate the application of the decoupling and control point position factors, we will use an example of an equipment provider.

Figure 3.4 represents a flow diagram of an equipment manufacturer. A flow diagram simply depicts the general relationships between resources and direction of flow of material (which moves from left to right). The bolded rounded box surrounding the flow diagram represents the boundaries of the system. Raw and purchased component stocks can be held on-site so they are depicted as being within the boundaries of the system.

This company generally pulls from raw stock held on-site. The first step is to cut material either through shearing, sawing, or by laser. The cutdown material is either sent to welding or to machining. From welding, material can go to assembly, directly to paint (service parts or options), or to an outside plating operation depicted by the dotted arrow going outside the system boundary to plate. Any plated work comes back to the company (the dotted arrow returning) and goes directly to assembly.

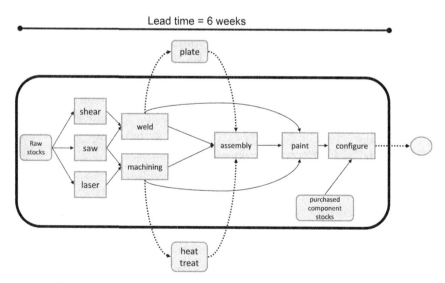

Figure 3.4 A Decoupling and Control Point Placement Example

From the machining area material can either go to assembly, directly to paint (service parts or options), or to an outside heat-treat facility. Once again, the dotted arrows represent the path to and back from the third party. Once received, the heat-treated items are sent directly to assembly.

Once assembled, the assembly goes through paint where it is painted a customer-designated color. From the paint operation, assemblies and options go to final configuration where they are combined with purchased components to meet the customer specification. Service parts can go directly from paint to a customer or to the assembly line as some service parts are also the exact same part number as some option parts (e.g., bucket teeth).

The lead time to consistently accomplish this process is currently six weeks. The customer expectation is to have a fully configured machine in two weeks. The Sales Organization has indicated that there are significant benefits to a one-week lead time. The team decides to go for a system design with a one-week lead time.

Given the lead-time objective, where should this company consider placing decoupling points? According to the operations staff, a one-week lead time would require a decoupling point in front of paint. Paint is a major convergent point, being fed by welding and machining for spares and options as well as assembly. A lot of variability from the preceding processes (including outside processors) can be absorbed at this convergent point. Going farther back in the process will extend the lead time beyond the one-week objective. Having a decoupling point between paint and final configuration would not make sense because the color can be unique to the customer.

The staff concludes that if assemblies, spares, and stores were always available to paint, and purchased components were always available for final configuration, then a one-week lead time could be a reality. They decide to place decoupling points accordingly. Figure 3.5 shows decoupling points (bucket icons) inserted on the paths coming into paint and at purchased component stocks. The lead time is indicated as one week from paint through final configuration. This is the lead time the market will experience.

The lead time feeding the decoupling point in front of paint has been compressed to three weeks after the team commits to a decoupling point at raw stocks, taking supplier lead time completely out of the internal planning and scheduling equation. Furthermore, the team believes that decoupling

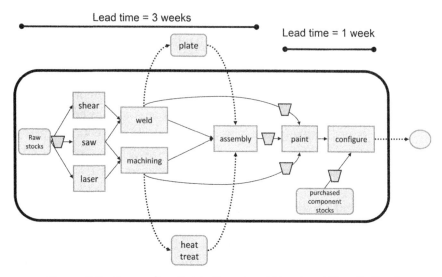

Figure 3.5 Decoupling Points Placed and New Lead Time Defined

the paint schedule from the schedules of assembly, weld, and machining will reduce the noise and conflicting signals in those areas. The reduction in variation in the feeding resources should translate to lead-time gains.

Now the team begins looking for control points. The team agrees that with the defined decoupling point positions, there are no serious capacity constraints. Selecting all or some of the entry and exit points can help protect the schedule against variability inherent in dealing with external entities on an order-by-order basis. The team concludes that with the new short lead-time promises, the ultimate control point is the scheduled delivery to a customer (an exit point). One concern the team has is the heat-treat processor. In the past there have been repeated challenges getting the work back within a reasonable time. The team admits that the processor has often been given jobs that are already late. They believe that placing control points both outbound and inbound will help stabilize the supplier's schedule as well as establish a much better ability to measure both their and the supplier's performance. The team has confidence in the plating third party's reliability. They believe that by sending the job out on time they will have no performance issue. They place a control point outbound but not inbound.

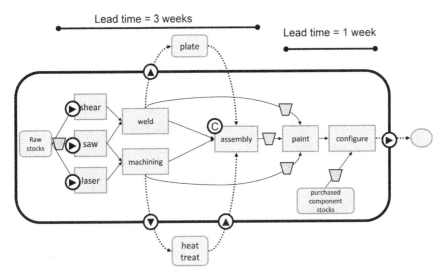

Figure 3.6 Decoupling and Control Points Illustrated

In Fig. 3.6 these control points are illustrated by circles with directional arrows on the dotted arrows leaving and/or returning to the system.

The team now turns its attention within the boundaries of the system and before the decoupling points in front of paint. How to synchronize this part of the plant? There is a point of convergence that seems to work well—assembly. The team reasons that if the schedule of assembly is finitely built and carefully managed, it will reduce the chaos in operations today, characterized by part stealing and lots of open orders with missing parts. They place a control point at assembly. In Fig. 3.6 this is depicted by a circle with a "C" inside of it.

As a convergent point, the assembly schedule will determine all other schedules in front of it including the schedules of the control points to outside processors. The assembly schedule is the primary control point for this portion of the plant—its schedule trumps all others. The team decides to also declare material release to the floor as a control point. This will ensure that work enters the system when it needs to in order to hit the assembly schedule at the appropriate time. This is indicated in Fig. 3.6 with separate circles with inbound arrows between raw stocks and shear, saw, and laser.

Protecting Decoupling and Control Points

At these decoupling and control points, we will have to employ some form of dampening mechanism to absorb accumulated variability so that these points can achieve their intended purposes of flow protection and control. This dampening mechanism is called a buffer. There are essentially three types of buffers to employ at these points: stock, time, and capacity.

Demand Driven Stock Buffers

Stock buffers are quantities of inventory designed to cushion a position or place against variability. There are many types of stock buffers in conventional operations, but many of them suffer from antiquated or incomplete rules or assumptions in light of the increased volatility and complexity of the New Normal. Through numerous examples and decades of experience, the authors advocate the stock buffering mechanisms in DDMRP methodology. These buffers are called strategically replenished buffers and should be the dampening mechanisms used at decoupling points.

These buffers are most often the heart of a demand driven system. Their importance will continue to grow as customer tolerance times continue to shrink, forcing decoupling points closer and closer to order fulfillment. These positions are primary planning and scheduling positions that launch new supply orders and convey priority for open supply orders based on buffer status. They are designed specifically to dampen in both directions (up and down the dependency structure), thus mitigating the bullwhip effect and protecting or promoting flow. Thus, in this chapter, more attention is paid to these types of buffers than the other buffering mechanisms.

Strategically replenished buffers are initially sized through a combination of factors including an average rate of use, lead time, variability, and order multiples. These buffers are stratified into colored zones (green, yellow, and red) for easy priority determination in planning and execution. Each zone has attributes that affect its relative size. The buffers dynamically adjust with market changes in consumption or in advance of planned or known activity such as seasonality or promotions. These buffers are carefully replenished based on what DDMRP terms an "available stock equation" and its relation to the yellow zone of the buffer. An available position stock in yellow creates a supply order against the position. Figure 3.7 illustrates the nature of these buffers.

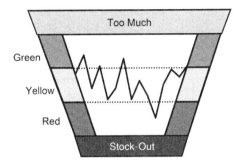

Figure 3.7 A Strategically Replenished Buffer

The planning methods of DDMRP represent major changes to the rules of formal planning. The imposition of decoupling points in product structure and the disconnecting of forecasted orders from supply order generation flies in the face of convention.

Readers should not confuse strategic replenishment buffers with MRP's safety stock. Safety stock does *not* decouple; it seeks only to compensate for variability assuming no decoupling or lead-time compression (i.e., a longer planning horizon). This makes safety stock a very inefficient type of dampening mechanism. Additionally, safety stock has mechanisms that can exacerbate the bullwhip effect.[5]

The specific sizing, management, and measurement of these buffers is detailed in Chapter 11 and Appendix A of this book as well as in Chapters 24–28 in the Orlicky text by Ptak and Smith.

Demand Driven Time Buffers

Control points are just that—points of control. The activity between decoupling points is typically managed through control points. Their schedules pace all other resource and area schedules. The protection of these schedules is crucial for overall control. Demand driven time buffers are planned amounts of time inserted in the product routing in order to cushion a control point schedule from disruption. Time buffers are sized based on the reliability of the string of resources feeding the control point.

---

[5]For more detailed description of the difference between safety stock and strategically replenished buffers, see the white paper, "Replenishment Positions vs. Safety Stock, Why are they so different?" by Carol Ptak and Chad Smith, Demand Driven Institute, 2011.

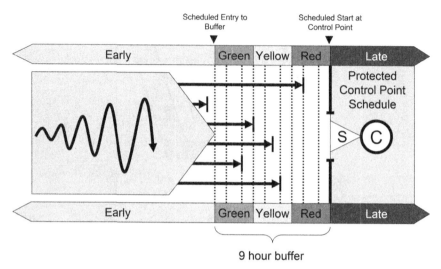

Figure 3.8   Time Buffering

The less reliable or more variable that string is, the larger the time buffer requirement.

Figure 3.8 illustrates the concept of the time buffer. The time buffer is in the middle. It is the range bordered on the top and bottom by boxes containing the words *green, yellow,* and *red.* On the leftside of the buffer is the flow of work from preceding operations towards the buffer. This flow of work is represented by the shaded pentagonal figure pointed at the buffer. The squiggly line represents the variability in the flow of that work.

On the rightside of the buffer is the control point indicated by the shaded box with a circle with a "C" inside of it. The triangle with an "S" inside of it is meant to indicate the scheduled start of work for an order at the control point. The buffer is divided into three zones titled green, yellow, and red. In this example, the total buffer is nine hours of time. Each zone has been set at three hours of duration. The dotted lines that bisect the buffer from top to bottom indicate each hour of each zone.

With a nine-hour buffer, work orders are scheduled to be in the buffer (buffer entry schedule) nine hours before their scheduled start time at the control point. With the existence of the variability in the preceding workflow, that will rarely happen. The most likely scenario, assuming the buffer is sized properly, is that the majority of work orders will arrive in the buffer sometime between the buffer entry schedule and the scheduled start of work at the control point. In Fig. 3.8, this is depicted through the various lengths of the arrows into the buffer. These are called buffer penetrations. A buffer penetration occurs when work is not in the buffer and available to the control point any time after the scheduled buffer entry time. In some cases, work actually arrives at the buffer before the scheduled entry in the buffer. This is called an early entry. There is one instance of that in the figure.

The lengths of these buffer penetrations will determine the risk to the control point schedule and whether expedite action is required in the preceding resources. The longer the penetration the larger the risk to the control point schedule. The key is that when the length of a penetration does go beyond the scheduled start of work at the control point—a late entry in the buffer will be created. A late penetration means that the control point schedule has been compromised.

Because these buffers are a part of a system's total lead-time equation, a company should constantly strive to reduce these buffers while limiting buffer penetrations in the red and late zones. Later in the book, we will discuss the importance of collecting data about these penetrations and their vital role in smart metrics.

Demand Driven Capacity Buffers

Capacity buffers protect control points and decoupling points by giving resources in the preceding workflow the surge capacity to "catch up" with variability. Thus, a capacity buffer is protective capacity that provides agility and flexibility. This can allow both stock and time buffers to safely be reduced while ensuring service level targets.

Figure 3.9 shows a resource's load requirements over 11 time periods. The black bars are meant to convey load—the longer the bars, the bigger the load. The capacity buffer is the section stratified by R, Y, G (standing for red, yellow, and green, respectively). The black bars in three of those time periods penetrate the buffer. The higher those bars go, the closer a resource gets to being overloaded in that period. A resource that is consistently loaded to red or overloaded is simply less responsive. It is becoming capacity constrained and should be considered for either control point status or capacity upgrades.

This protective capacity is often available today in every resource that is not capacity constrained. This protective capacity *should not* be used to

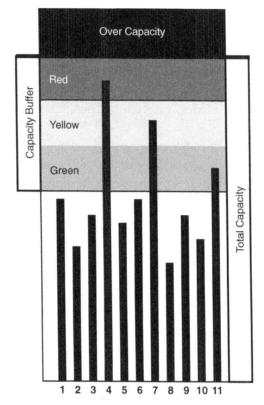

Figure 3.9 A Capacity Buffer

improve unit cost or to drive a particular resource's utilization. In fact, the entire notion of a capacity buffer flies in the face of conventional costing policies. Capacity buffers require that a resource maintain a bank of capacity that goes unused. Exploring ways to minimize the investment and expense associated with this unused capacity is totally valid. What is *not* valid is encouraging a resource to *misuse* its spare capacity to improve unit cost or resource efficiencies by running unnecessarily. When that happens, responsiveness goes down and the stock and time buffers will be jeopardized, forcing these buffers to be increased to compensate. This will result in increased lead times and inventory levels throughout the system.

Returning to our previous equipment manufacturer example allows us to see the completed design. Figure 3.10 illustrates that completed design. Buffers have now been inserted or depicted. The bucket icons depicting

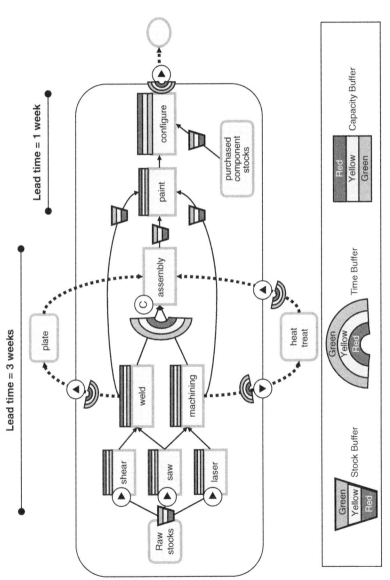

Figure 3.10 The Completed Demand Driven Design

strategically replenished buffers have the green, yellow, and red stratification. The radial green, yellow, red icons represent time buffers in front of the control points.

All resources that are not control points (all resources except assembly) will have capacity buffers. These buffers are represented with a stratified box in the top portion of the resource box. It is important to understand that these capacity buffers are not meant to convey that the organization simply plans to invest in capacity everywhere. It does mean that the company will commit to keep more capacity in those areas relative to the finitely scheduled control point that those areas feed.

Bringing the Demand Driven Model to the Organization

The people in the organization have powerful intuition—it just needs to be harvested and directed. It never ceases to amaze the authors how easily the people that perform the work, that crank the wheels, push the buttons, build and manage the schedules, and referee the chaos grasp the concepts—it makes sense to them. They simply got used to living with and complying with the restrictions of convention (unit cost focused metrics).

When people can see and effectively communicate how the system works while understanding its objective, dramatic things can happen and they can happen quickly. The authors are reminded of a situation that once happened on the floor of a printing company. The company was under competitive threat. Manufacturing responsiveness had to increase dramatically.

Lead times needed to shrink and due dates needed to be much more reliable. Those were prerequisites for defending market share but, more importantly, to support a new type of offering that would make a huge impact to total sales. Replenishment buffers over a wide array of products would be placed at the company's factory warehouse. Additionally, consigned stocks would be positioned at remote customer locations and replenished as needed pulling against the plant's finished stock buffer.

Management was worried that without increased responsiveness in the plant:

1. The finished stocks necessary would be too large for the space and cash available.
2. The plant would not be able to keep up if the offer took off (they were convinced it would be very well received).

The problem with getting the plant more responsive resided at the laminator; it would not be able to keep up with demand and the resources that fed it and the resources that took from it. It would be the control point while a new facility with a new or additional laminator could be brought online. In the front office, a plan was devised to protect the laminator's capacity as much as possible through finite capacity scheduling, the use of a time buffer and larger combined process batches, and a third shift (something that management knew would be expensive, unpopular, and inefficient). Figure 3.11 represents their demand driven design.

An all-company meeting was called to lay out the plan and communicate the need to implement this new system. The general manager, Kirk, carefully went through the logic, doing an excellent job of explaining the need for change. "We are under a real threat here. We have an idea for a market offer around service that we believe will lead the industry, but . . . it requires us to be a much more responsive manufacturer."

Then the lights were dimmed and the entire company watched the movie version of *The Goal.* When the lights came back on, the general manager displayed the solution design, and explained the offer and why the market should appreciate it. Then he began to explain what had to change inside the plant to support that offer, clearly marking the laminator as the control point. "Our current capabilities will have difficulty supporting this offer because the laminator cannot keep up. We need to make changes here while we bring additional capacity online." A hand went up in the audience—one of the laminator operators.

"Yes, you have a question," the general manager said, slightly annoyed at the disruption to his groove.

"Ah, yes," the laminator operator said. "So you need the laminator to run faster, right?"

"Yes, as I said before, it's the constraint," replied the general manager. "It simply does not have the same amount of capacity in relation to the other resources. That's why we need to do something in the short term about better managing its capacity. In the long term we will need to get more lamination capacity."

"Now, as I was saying we are first going to ask for any volunteers from existing employees that may want to work a graveyard shift. We know that we need to hire . . ." Again, a hand goes up from the same operator who asked the first question.

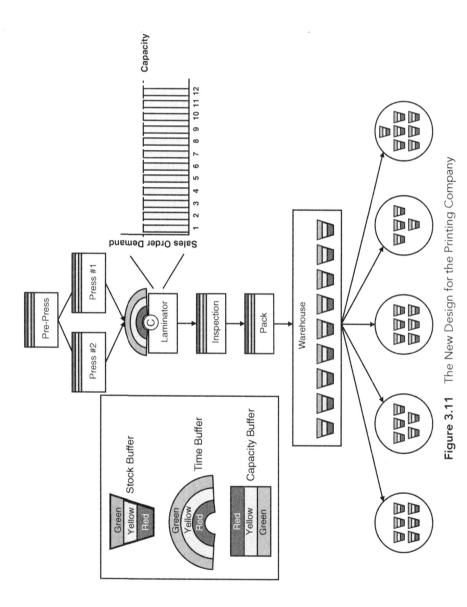

Figure 3.11 The New Design for the Printing Company

"Yes, you have another question?" asked the general manager.

"Well, not really a question, more of a comment really," replied the operator.

"OK, everyone's opinion is welcome; we know this is going to be a big change for many of you," replied the general manager.

"Thank you, I appreciate that. I would just like to point out that the laminator can run quite a bit faster than it currently does," said the operator.

The general manager looked at him funny. "And how is that? In the two years I've been here it has never gotten more than 3 million sheets per month out."

"Well, it has a rate setting dial. Right now the dial is set to '5' and it goes up to '10.' We have never changed the setting on it in the 10 years I have worked here. The only reason I even know about the setting dial is that I had to make a small repair to a belt a few years ago. The dial is kind of hidden. I told Bill (the previous plant manager) about it. Maybe we could try gradually increasing the dial and see what happens?"

The general manager was visibly stunned. "Can you show me?"

"Sure come over to the laminator, like I said, it's kind of hard to see."

The dial was turned up and the rest is history. The company did not buy a new laminator or move to another facility or add another shift. They did implement the new scheduling system around a new constraint in pre-press and successfully deployed the new market offer. The results in the first two years:

Inventory Turns: 5 to 16 per year
Total Inventory: 3 times the number SKUs with 30 percent less inventory
Lead Time: 16 days to 3.2 days
Order Fulfillment: 70 percent to 98 percent

The point of this story is not to make the general manager or the front office look incompetent. They had the right direction for sales. They were thinking flow. But they, along with the advice of their consultants (the authors of this book), were about to make a huge commitment and change relying on one critical false assumption; an assumption that was quickly blown out of the water by the resource operator. For the general manager and the consultants it was embarrassing *and* fulfilling at the same time. The point of this story is to illustrate that the people in our organizations have incredible knowledge and intuition. When given the proper framework, context, and opportunity, they can make a huge impact on a system's flow and ROI in a very short period of time. That is what the use of thoughtware is all about.

Operating and Sustaining the Demand Driven Model

While sharing the demand driven model with the organization can lead to some quick and powerful insights, even a new design, the key to getting people to take actions to sustain and improve within it resides in our ability to deploy metrics that convey clear, appropriate and visible signals to make that happen. There are quick fixes—they just have nothing to do with managing costs and everything to do with promoting visibility to the flow of both information and materials to and through the decoupling and control points.

Management can't facilitate ROI effectively if it doesn't understand flow in the context of the system. You can't build metrics that encourage flow if you don't understand flow in the context of your resource base and the market. Now that we have discussed these topics, at least at a cursory level, we will turn our attention to building the new metrics. We will have to continue to chip away and reinforce the concepts in the first three chapters, as the problem is so pervasive and ingrained it continually places an organizational "fogbank" on what really matters. You simply can't measure what you can't see. You can't see what you don't define or, worse, obscure.

Russell Ackoff, a famous pioneer in the field of operations research, systems thinking and management science expressed it best in his book *A Little Book of f-laws - 13 Common Management Sins:* "Managers who don't know how to measure what they want, settle for wanting what they can measure."[6]

The rest of this book is dedicated to the removal of the fogbank and conveying clear signals to our organization in order to operate a demand driven system in the New Normal.

---

[6]Russel Ackoff and Hubbert Addison, (Triachy Press, 2006, page 4)

CHAPTER 4

Introducing Smarter Metrics

The first three chapters of this book have provided a cursory overview of some fundamental breakdowns in the way that most companies operate today in light of the circumstances of today. Those breakdowns range from the way that organizations plan and measure to the way they approach problem-solving. We propose an alternative: a flow-centric paradigm that logically and simply connects directly to return on investment. Furthermore, a method of operating within that paradigm has been revealed—position and pull. To institute that method of operation, an organization must take steps to become demand driven. Is it enough? For most companies, indeed, most of industry, the answer is simply, no. What has been presented so far is insufficient to reveal a deeper truth, a truth that, once revealed, simply changes everything.

The Search for a Deeper Truth

Today's deep truth drive industry to report on the efficiency of each resource, assuming that local resource efficiency translates to and drives total system efficiency. A tenth of an hour saved here saves a tenth of an hour for the whole system. That savings is quantified as a total cost savings defined by the sum of the unit-cost savings. The assumption is that the sum of the cost savings will fall to the bottom line. This deep truth is embedded in the product cost roll-up structure of every supply chain. In fact, manufacturing information systems with time standards and material requirements intended to plan, schedule, and execute have been transformed into product cost-centric systems to satisfy GAAP financial statement presentation for external reporting. Companies have lost their connection to flow over time and replaced it with a mathematically inappropriate equation of unit cost over time.

Today's deep truth leads companies to operate as if the first law of manufacturing connects all benefits directly to the minimization of unit cost (push and promote) *not* better flow performance. This drives all reporting and measures and tactical planning and execution actions to the following objectives:

- ▲ Minimize total product unit cost;
- ▲ Maximize resource utilization;
- ▲ Strive for positive overhead, labor, and volume variances;
- ▲ Initiate cost-reduction efforts with emphasis on machine, labor, and inventory reductions that quantify the expected savings on fully absorbed standard cost.

Replacing today's cost-centric version of the first law with the proper flow-centric version will require exposing, understanding and adopting a deeper truth. The case and proof that cost-centric assumptions no longer make sense in the presence of new circumstances or evidence will have to be extremely compelling.

What has been presented so far in the first three chapters is not nearly compelling enough. The rest of this book is dedicated to making this case and proving its effectiveness and sustainability. And let's not forget that negating a deep truth is a double-edged sword. The good news is that the new information gives us an updated and more relevant way of thinking that allows for huge gains. The downside is that entire organizations, infrastructures, and careers have been built upon the false assumptions of unit cost minimization.

The case for a deeper truth began in Chap. 1 with the discussion of Plossl's first law of manufacturing, connecting the rate of flow of materials and information directly to ROI. If embraced, this should drive all reporting, measures, tactical planning, and execution actions to the following objectives:

- ▲ Synchronize demand and supply signals between the defined critical points in flow;
- ▲ Prioritize improvement efforts based on identifying and then removing whatever blocks flow;
- ▲ Quantify the results of the above actions by the net change in the revenue and cost side of the ROI equation as well as the net change on the investment side of the equation.

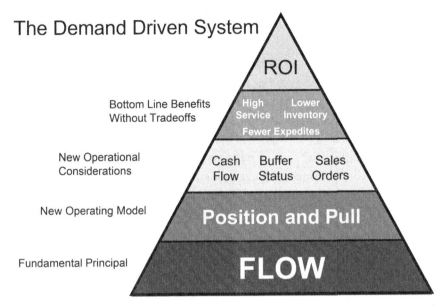

The Demand Driven System

ROI

Bottom Line Benefits
Without Tradeoffs

High Service Lower Inventory
Fewer Expedites

New Operational
Considerations

Cash Flow Buffer Status Sales Orders

New Operating Model

Position and Pull

Fundamental Principal

FLOW

Figure 4.1 The Demand Driven Materials Requirements Planning (DDMRP) Pyramid

Figure 4.1 depicts this new way to operate whose very foundation lies in embracing the concept of flow *not* cost. Unfortunately, simply starting at flow does not dive deep enough. How much deeper must we go?

Quantifying the Performance Gap

Part of understanding why a change is required is knowing what opportunities are currently being missed by not changing. It is possible to quantify the gap between the cost-centric world of push and promote and flow-centric world of position and pull. The formula in Fig. 4.2 expresses the gap between the two worlds. It quantifies the potential system improvement

$$\Delta \text{Visibility} \rightarrow \Delta \text{Variability} \rightarrow \Delta \text{Flow} \rightarrow \Delta \text{Cash Velocity} \rightarrow \Delta \left(\frac{\text{Net Profit}}{\text{Investment}} \right) \rightarrow \Delta \text{ROI}$$

Figure 4.2 The Gap Formula between Push and Promote and Position and Pull

or degradation in moving from one world to the other, with the following points:

▲ Visibility is defined as relevant information for decision-making;
▲ Variation is defined as the summation of the differences between what we plan to have happen and what happens;
▲ Flow is the rate in which a system converts material to product required by a customer;
▲ Cash velocity is the rate of net cash generation (sales dollar – truly variable costs) – operating expense otherwise known as throughput dollars – operating expense dollars;
▲ Net profit/investment is of course the equation for ROI.

Note that this formula starts not at flow but at what makes information relevant. If we don't fundamentally grasp how to generate and use relevant information, then we cannot operate to flow. Moreover, if we are actively blocked from generating or using relevant information then even if people understood there was a problem, they would be powerless to do anything about it. Thus, we have reached the core problem plaguing most manufacturers today, *the inability to generate and use relevant information to drive ROI.* Without addressing this core problem, there can be no access to deeper truth.

Figure 4.3 illustrates the delineation between what Plossl and Chaps. 1–3 highlight versus the core problem area that must be explored for a deeper truth. The rest of this book will be devoted to an in-depth exploration of the core problem area and its implications, in order to identify and deploy smarter metrics.

Smarter Metrics

Smart metrics do not put resources in conflict with themselves or the rest of the organization. They focus on the visibility of relevant information, in a relevant time frame, to direct action and align priorities across

Figure 4.3 The Area of Deeper Truth versus Plossl's First Law

the organization. They focus on the rate of flow through the system and identify what is blocking it and the actions to remove it. They focus on system stability and reliability that result in high due date performance. Smart metrics produce visible, easy-to-understand signals to reduce contention and eliminate conflicts for resources and material. Smart metrics are system-flow based *not* unit-cost based. They are *not* a source of variation; that would simply not be smart.

Below is a summary of four companies who agreed to share their results. Their results are typical of companies who can break out of the core problem, identify what is relevant, and build a systemic demand driven approach. They have all sustained and built on their initial improvement results continuously year after year. They have successively driven ROI improvement through industry upturns, downturns, and the most recent recession, and they have used smart metrics to do it.

Company 1: Oregon Freeze Dry

With six plants in three countries, Oregon Freeze Dry, Inc. (www.ofd.com) and its affiliate, European Freeze Dry, is a global leader in freeze-drying technology. It is the world's largest diversified freeze dryer.

Business Challenge

- ⚠ Late shipments
- ⚠ Retail stock-outs
- ⚠ Lost sales opportunity
- ⚠ High expedite-related waste: overtime, late shipments, premium freight

The Results

Division A: Mountain House

Sales: Up over 20 percent

Shipping: July–October 1997 (prior to DDMRP)   1164 shipments; 245 shipped late

=July–October 1998 (after DDMRP)   1697 shipments; 11 shipped late

Inventory: 60 percent less inventory on hand

Division B: Ingredient Business

Lead Time: 10–12 week lead time reduced to 3 weeks
            60 percent reduction in make-to-order lead time
Shipping: Improved from 97 percent to 100 percent on-time delivery
Inventory: Reduced by 21 percent

Since the beginning of their journey in 1998, they have experienced a 13× growth in sales and only a 2× growth in inventory with no major capital additions, all while maintaining greater than 98 percent on-time delivery.

Company 2: Jamestown Container Company

Jamestown Container Company (www.jamestowncontainer.com) is a large regional packaging solutions company headquartered in Western New York. They supply custom box and plastic packaging, litho laminated containers, wholesale shipping supplies, contract packaging services, and product package design.

Business Challenge

Jamestown Container competes in an industry mired in overcapacity, high fixed costs, intense price competition, and spiraling margin erosion. Jamestown Container was looking for a solution that would create a unique value for their customers through shorter lead times and higher customer-service levels. Unparalleled speed, quality, and service were core to their strategy.

The Results

▲ Operating profit up 300 percent
▲ Inventory down 40 percent
▲ Lead times down 70 percent
▲ Expedites nearly eliminated
▲ Inventory turns from 10 to 42

Company 3: LeTourneau Technologies, Inc.

Dubbed the "mighty movers of the earth," LeTourneau Technologies, Inc. (www.joyglobal.com) is a global group of best-in-class organizations specializing in the design, manufacture, implementation, and effective use of advanced technologies for onshore and offshore oil and gas drilling, forestry, mining, and steel markets. In 2011 LeTourneau Technologies was acquired by Joy Global. The LeTourneau Technologies story is referenced in the foreword by Dan Eckermann, the former CEO of LeTourneau Technologies.

Business Challenge

In a 2008 presentation at the Constraints Management User Conference the head of LeTourneau's Strategic Planning and Logistics function presented the following analysis and results.

"MRP was simply not working in the face of our supply chain variability and volatility."

- ▲ MRP was pushing, they wanted to move to customer pull
- ▲ Material shortages causing cascading disruptions across the company
- ▲ Simultaneously experiencing chronic shortages and carrying too much inventory
- ▲ High overtime and expediting
- ▲ On-time delivery around 70 percent

The Results

Implemented demand driven principles in 2005:

- ▲ 6× growth in sales revenue with only 80 percent growth in inventory dollars
- ▲ ROI from 4 percent to 22 percent
- ▲ Lead times for loaders cut in half
- ▲ Lead times for new drilling rigs cut in half
- ▲ Steel mill lead times from 14 weeks to 4 weeks

Company 4: M.C. Gill Corporation

M.C. Gill (www.mcgillcorp.com) is one of the world's largest manufacturers of honeycomb, high-performance floor panels, cargo compartment liners, and original equipment for passenger and freighter aircraft.

Business Challenge

How to deal with the inherent complexity and volatility in the material, repair, and overhaul (MRO) market? M.C. Gill wanted to increase sales of

MRO products and still minimize their inventory investment. The ability to deliver the same day to customers with an AOG (airplane on ground) was the market goal. Relying on forecast and safety stock resulted in large inventories but chronic stock-outs.

The Results

Sales of MRO products increased by 10 percent over the prior year but the profit percentage impact was dramatically higher. Most importantly, they had a same-day fill rate of 100 percent despite the dynamic marketplace. After implementing their demand driven model, they quickly reduced the number of items that fell below the minimum service target level. The numbers of items below the minimum strategic stock level continued to drop, and stock-outs are virtually nonexistent. Inventory investment dropped 5 percent while service levels soared and expedite costs disappeared.

Lessons from the Four Companies

Each of these companies emerged from the 2009–2010 recession stronger and with more market share. They represent widely different markets, product complexity, geographic locations, technologies, and cost structures. Yet all of them applied the same basic approach to managing their supply chain and reaped very similar results. This underscores the potential for more companies to make similar leaps forward in both due date performance (DDP) and ROI by following the same steps pioneered by these companies. In fact, the authors believe that to thrive in the New Normal, there is no alternative.

Remember, the only real proof that something is the right thing to do is to confirm its connection to an improvement in ROI—the system's ultimate measure. When *all* of the key indicators of ROI go the *right* way *together* there is a dramatic improvement in ROI. A demand driven strategy often ensures all of the KPIs underpinning ROI go the right way together. The proof is in the results of the companies who have changed how they think, what they measure, and the actions they take on a daily basis. They emerged on the other side of the gap formula in Fig. 4.2 with dramatic improvements—improvements that started with an ability to define what relevant information is.

The change necessary to negate a deep truth requires more than proof it works in other environments; that is simply a recipe to hear excuses about how different those environments must be. Instead we must address the core problem identified previously in this chapter. That core problem, once again, is the inability to generate and use relevant information to drive ROI. Chapter 5 will use an example to explore the depths of the core problem.

CHAPTER 5

How Do We Know What's True?

Everyday managers throughout an organization are asked to make decisions and take actions to both plan and execute their organization's strategy. Often their decisions and actions are contrary and/or negated by the decisions, actions, and priorities of their peers. They also find themselves taking the opposite action the next hour, day, week, or month. Understanding and addressing the source of the confusion and the ever-changing priorities begin with understanding the cause and the cost to the organization. If a business has a coherent strategy, then all parts of the organization should agree on the planning and execution priorities. For the parts of the system to stay coherent during planning and execution, all of the parts must understand the predictable outcomes from each other's decisions and actions.

All of the parts need the same definition of *relevant information* and the same view of the *state of the system* to make decisions and take actions that keep flow priorities aligned. If not, the organization will continue to firefight the same old symptoms and generate the same suboptimal financial results. How does an organization know what is true when its managers from top to bottom do not see and cannot agree on the same picture of reality?

Chronic and pervasive conflicts in organizations compromise performance. These conflicts have been around so long they have simply become a part of the environment; a given. In other words, people are blind to them. Understanding these chronic conflicts and their respective root causes or assumptions can help explain and link how many of the dysfunctional actions people take are directly related to the measures that are imposed on them.

The ability to turn the process of resolving these conflicts into a team search for solutions rather than a witch hunt for whom to blame is critical to building a one-team approach. Chapter 2 introduced the installation of thoughtware, people's ability to think and communicate systemically to help identify and resolve systemic conflicts and oscillations. Instead of criticizing each other's decisions and actions, the focus is to identify and eliminate the policies, work practices, or measures that consistently drive a bad decision, and the corresponding actions that erode flow and financial performance. After identifying the core problem, we need to answer the following questions to identify a systemic solution:

1. Where do the measures, work practices, and policies come from? What was their original objective? What are they designed to protect?
2. Are the reasons above still relevant in today's New Normal?

Although they are still common in organizations, many core strategic policies, work practices, and measures are no longer relevant in our current world. They are obsolete due to changes in consumer expectations, technology, market structure, and/or scientific breakthroughs. Chapter 1 provided a brief exploration of two critical areas, planning and costing, in which we see this effect. These two critical areas are at the heart of the core problem identified previously in Chapter 4. Today these areas are unable to generate and use relevant information and the unit cost inormation used is the source of tremendous conflict.

To prove this point we will take a fictitious (but very realistic) company story to begin to expose this conflict. The name of this fictitious company is simply "Company Normal." It will be our vehicle to demonstrate how the four-step scientific method below can be applied to a complex situation. In general, the scientific method describes a sequence of steps that must be followed if an idea is to be accepted in the scientific community. The method allows for, and expects new information. Over time new discoveries are assimilated into the existing canon of ideas or beliefs. That is how science evolves. Let's see if it can bring certain things to light at Company Normal.

The Scientific Method

1. We see something unexplained.
2. We develop an explanation.
3. We test the explanation with an experiment that gives us facts.
4. We evaluate the facts.
   a. If the facts support the explanation, we have a theory.
   b. If they don't, we need to go back to Step 2, change our explanation, and repeat the process.

The first step in the scientific method is the observation of something unexplained. We can observe Company Normal's monthly meeting. In attendance are the sales manager, plant manager, accounting manager, and engineering manager.

Monthly Meeting of "Company Normal"

The **sales manager** reviews the backlog of unshipped orders, then raises his concern that the plant manager is not focused on the right priorities. He is not managing his plant effectively, as evidenced by both the late orders and the low fill rates on make-to-stock items. There are stock-outs on products they have committed to hold for customers and late shipments on make-to-order products. The plant manager appeals to an independent source—the accounting manager—for confirmation, the proof he is managing his plant efficiently.

The **accounting manager** supplies the reports that clearly show all of the plant resources' utilization and efficiency are in the high 90's. Last month the plant produced parts at a rate of volume that resulted in a favorable absorption variance, and the plant is ahead of the operating plan for the year. Accounting also points out it is the plant manager's ability to continually beat his efficiency standards and lower the unit product costs that has kept product margins from eroding even further. Accounting produces another report demonstrating sales' continual price concessions and the resulting product margin erosion.

The **plant manager** seizes the opportunity to point out they need more capacity. He is running the plant flat out and can't keep up, as evidenced by the backlog of unshipped orders and the product stock-outs. His budget for overtime is maxed out from all of the expediting to try to address the stock-outs and the late sales orders.

The **accounting manager** confirms the overtime expense is at an all-time high and so is expedite freight charges. They both need to be better controlled.

The **sales manager** states the plant must do whatever is necessary to deliver on time and get the lead times down. Work more overtime, and if the plant is really out of capacity then there appears to be no choice but to make some capital investments or hire more people. He states the risk of losing customers if service and lead times do not improve is real—they lost two accounts last month due to service issues and despite price concessions.

The **accounting manager** points out they will have to borrow the money because cash flow is really tight right now, even in spite of last month's record profit. He then moves onto the topic he is most concerned about: sky-high inventory and low turns. What can be done to get inventory down? Invested dollars in work-in-process, subassembly parts, and finished-goods inventory are all up.

The **plant manager** points out that the plant is on target to their operations plan, but sales are not. Obviously, Sales needs to sell more and they are not selling the right products. If sales don't increase, the plant will not be able to continue meeting the production plan—there will simply not be enough warehouse space to store the unsold inventory. They have even filled up the off-site warehouse space they rented to store the slow-moving inventory. He suggests Sales should sell and promote the slow-moving items.

The **accounting manager** rolls his eyes and comments. "If you discount the slow-moving products in inventory, our product margins will erode further. Obviously we need to step up the cost-savings programs."

The **sales manager** points out they have orders and they would have sales if the plant could ship on time and they weren't stocked out and losing customers. He also points out that the slow-moving items represent more than six months of potential demand. Why doesn't the plant make what the market wants now?

The **plant manager** ignores the comment and points out they have no room for the new equipment they may need to expand the plant. Perhaps the addition should include warehouse space. They can pay for it from the savings in rent for the off-site warehouse.

The **accounting manager** asks the engineering manager to prepare a capital asset request and reminds him corporate is not approving any requests that do not have a payback of 18 months or less—cash is tight. "With our current ROI, we will not get any capital approved unless the case is compelling."

The **engineering manager** replies there should be no problem with justification. There are some new high-speed mixers that can double a production batch size in the same time as their current process and dramatically improve performance and deliver even lower unit costs.

They all turn to the next agenda item regarding the progress on the current cost-saving initiatives and pretend to ignore the elephant in the room. After all, they are used to the discussion. They have this meeting or some variant at the end of every month.

Walking out of the room, the plant manager asks the sales manager what accounts they have lost and who they are in danger of losing next. It now becomes clear to the plant manager that he is going to take some heat from at least one of the vice presidents, and maybe even the CEO, and very soon. Now he begins to worry. He tells the sales manager he understands and will take some actions when he returns to the plant to crank up the pressure on those orders even if he has to break into set-ups, set other orders aside, and rob the raw materials from other jobs. They will work Sundays if necessary. The sales manager thanks him and both hurry out to make phone calls to change priorities.

Using Consistent Terminology

The ideas in this book are challenging core beliefs—deep truth—and undoubtedly some emotional, defensive, and passionate debates will be triggered. Questioning someone's judgment can become a hot topic. In order to move beyond emotional arguments of the past, we need a consistent way of evaluating where speculation ends and evidence begins. Not to prove someone wrong, but to prove what policies and metrics need to change to stop the conflicting behavior and the waste it causes everyone. To get started, we need a common ground on terminology definition. What is the difference between a fact and a theory? How much evidence does it take to replace an existing theory with a new one? To answer these questions and make sure we're talking apples to apples, we need to clarify the words that are often used to justify various assumptions—words such as *science, fact, theory,* and *proof.* Because the ideas we are promoting are based in the sciences, we will stick to the definitions commonly used in the sciences.

What Is a Scientific Fact?

Scientific fact—A truth known by actual experience or observation.[1]

---

[1] www.dictionary.com

A fact is something having a real, demonstrable existence. For example, if the sales manager asks the plant manager for the status of a specific late order, the plant manager can provide a *fact* that the order is at a precise point in completion, at a precise location in the plant, at a precise time on a precise day. If the sales manager then sends his administrative assistant into the plant, and he confirms that the product order is indeed exactly where stated and in the condition of completion described by the plant manager, then the *fact* has been verified; and it was done by an objective witness who does not benefit one way or the other if it is actually not there. The fact tells us the current state, but it does not explain how the current state came to be. In other words, the fact does not describe when or how the order actually became late, although we can make an assumption based on the fact.

What Is a Theory?

> **Theory**—A proposed explanation whose status is still conjectural and subject to experimentation, in contrast to well-established propositions that are regarded as reporting matters of actual fact. *Synonyms:* idea, notion hypothesis, postulate. *Antonyms:* practice, verification, corroboration, substantiation.[2]

A theory is formed on the bases of facts that are known at the time. For the previous example, because the order for the product is late, it's reasonable for Sales to assume Manufacturing is not doing a good job of managing its priorities. This is an observed fact because the order was submitted inside the standard lead time for the product. This is the sales manager's *theory* of how the order became late. His theory is still conjecture based on facts but still a conjecture. It can remain a theory, and may even be a good one, as long as there is no evidence to prove it wrong. A theory can be significantly altered based on additional facts and new evidence over time.

What Is Proof?

> **Proof**—Evidence sufficient to establish a thing as true, or to produce belief in its truth.[3]

---

[2]Ibid.
[3]Ibid.

What is true in our example above? *Proof* is the evidence or argument that compels the mind to accept an assertion as true.

There is no use of the scientific method in Company Normal's dialogue. Facts, reports, and information are used to support arguments. The arguments are compelling because everyone believes they are using accurate and relevant information to make their decisions and to support their theories. Most of it is very accurate from a mathematical equation perspective, but it may not be relevant. The entire system of thought, reporting, and measures is based on a management strategy of maximizing efficiency and least cost, based on the mathematical model of full absorption cost accounting as defined by GAAP. In fact, every argument raised and every question that would challenge the status quo is diffused by this deep truth. We are not going to dispute the mathematical accuracy of the information, but we are going to question its relevance.

These are facts:

1. The plant has low on-time delivery performance for make-to-order product.
2. The plant has a backlog of make-to-order product.
3. There are stock-outs for make-to-stock product, and poor customer fill rates.
4. There is excess inventory in all forms; inventory turns are low.
5. They have poor cash flow.
6. They are losing customers.
7. Sales is missing plan.
8. ROI is low.
9. Resource efficiencies are high.
10. Unit costs are low.
11. Operations is making plan.

One of the theories being proposed on why the facts exist is the plant is not focused on the right priorities. This is easily dismissed by the report showing all of the resources the plant operates are efficient. The plant is beating the standard cost and producing a net favorable cost variance. Because of these high efficiencies, they are making the operating plan and beating budget.

Conclusion: The plant is efficient and focused on the right priorities. The evidence of this argument compels everyone in the meeting to agree even if it is not supported by the facts. Remember, the fact tells us what *is* but it may not explain how things came to be as they are. The elephant in the room is everyone knows there are unanswered questions—the questions remain unexplained because the team members can't verbalize a challenge to the deep truth of their strategy.

Remember the definition of *proof* is the evidence or argument that compels the mind to accept an assertion as true. The inability to challenge the conclusion despite overwhelming evidence that the facts do not support the conclusion is the indication of a deep truth. It can only be overcome with an even deeper truth.

What Is Scientific Proof?

Scientific proof—Based upon the pervious definitions, scientific proof is the proof that comes from facts as a result of scientific methods of discovery.

Let's approach the Company Normal example using a scientific proof and examine it using the scientific method.

1. We see combinations of fact that are unexplained:
   ▲ There is a large sales backlog but lots of excess inventory.
   ▲ The plant is capacity constrained but the warehouses are full of finished stock.
   ▲ The plant is on plan, sales are not . . . but the plant's plan was based on and derived from the sales plan.
   ▲ Net profit is on target but there is no cash available.
   ▲ We need more efficient equipment to be able to satisfy the market.
   ▲ We also need more warehouse space to store the product we are making that the market is not pulling.
2. We develop an explanation:
   ▲ The plant's focus on its monthly profit through cost of goods sold (CoGS) dollar credits is getting in the way of its ability to service the market.
3. We test the explanation with an experiment that gives us facts.
   ▲ We use rigorous logic to create solid effect, cause, and effect connections from the explanation as the core cause, to their facts resulting in their poor performance.

Our logical test uses a derivative logic process. A derivative process is a sequence of statements (as in logic or mathematics) showing that a result is a necessary consequence of previously accepted statements. The logic diagram in Fig. 5.1 is read from the bottom up with the tail of the arrow being "if" and the tip of the arrow being "then." The ellipse shape connecting the arrows is the logical "and," implying both conditions must exist to result in the cause at the tip of the arrow; either condition alone is not sufficient to create the cause. The example in Fig. 5.1 would be read as follows:

▲ "If" a key plant metric is to maximize their cost of goods sold dollar credit (the plant receives a credit value of the product unit cost for every product produced) "and" people behave according to their metrics, "then" plants try to maximize making high CoGs dollar products (products with high unit cost).

▲ "If" plants try to maximize making high CoGs dollar products, "and" some items have more CoGs dollars than others, "then" departments tend to prioritize producing the high CoGs dollar items at the expense of the low items (e.g., cherry-pick orders based on CoGs dollar value).

▲ "If" departments tend to prioritize producing the high "CoGs" dollar items at the expense of the low items (e.g., cherry-pick orders based on CoGs dollar value), "then" some make-to-stock products incur stock-outs.

▲ "If" departments tend to prioritize producing the high CoGs dollar items at the expense of the low items (e.g., cherry-pick orders based on CoGs dollar value), "then" some make-to-stock products are overstocked.

▲ "If" plants try to maximize making high CoGs dollar products, "and" plants receive the CoGs dollar credit when they ship to their own warehouse, "then" plants tend to produce to stock even when there is no demand signal (e.g., extend the forecast).

▲ "If" plants try to maximize making high CoGs dollar products, "and" setting up more increases product unit cost/lowers efficiencies, "then" plants pull ahead orders to increase the batch sizes for make-to-stock orders.

▲ "If" plants pull ahead orders to increase the batch sizes for make-to-stock orders, "and" plants try to maximize making high CoGs dollar products, "then" some make-to-stock products are overstocked.

▲ "If" plants pull ahead orders to increase the batch sizes for make-to-stock orders, "and" make-to-stock and make-to-order share common capacity and materials, "then" make-to-order backlogs grow—we ship late.

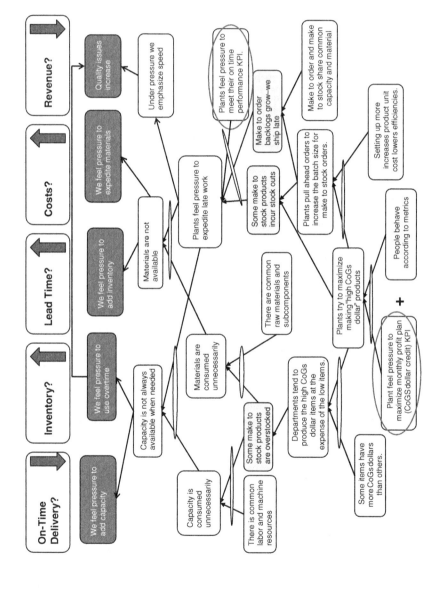

Figure 5.1 A Logic Construct to Identifya Core Problem for Company Normal

Figure 5.1 connect show the plant's actions to meet their planned absorption rate measure. CoGs credits in the diagram become the negative effects (shaded boxes in the middle of the diagram). They are then forced to take firefighting actions when they feel pressure to meet their other KPI of maximizing due date performance. What is the predictable effect of the actions taken to meet customers' delivery due dates and their fill rates for make-to-stock product? The diagram shows us: all of the key performance indicators of ROI are going the wrong way together.

The actions taken to meet their absorption cost measures, actually *light the fires* they end up fighting with the actions necessary to try to meet their customer service measures.

The central warehouse and sales organization put tremendous pressure on the plant to expedite the stock-outs as well as their late customer orders when the results become too painful. Yet the plant is also expected to meet their "budgeted or planned" shipment dollars (or tons or average gross margin or whatever standard cost-centric efficiency measure the organization employs). This is regardless of the fact that the original sales plan, on which the plant's operating budget and absorption rate were calculated, has nothing to do with what the market is actually pulling. Note again, the shaded boxes at the top of Fig. 5.1 are the undesirable effects or symptoms the client identified in the original negative effects listed above. The actions they take create the negatives they are working so hard to overcome. There is no way off the hamster wheel of blame. The net effect:

▲ The plant is forced into taking actions that oscillate between attempting to meet the target efficiency/plant performance measure and then scrambling to recover their on-time delivery to their customers.
▲ The sales manager is forced into taking actions to reduce prices and run extra promotions to compensate for the poor service levels.
▲ The engineering manager is forced into justifying capital improvements that will increase individual batch sizes and make the inventory, service levels, and expedite costs even worse.

The oscillation of cost- and service-centric behavior creates tremendous resource and spending waste, erodes due date performance, and increases investment. ROI goes in the wrong direction.

The majority of negative effects on the "fact list" are directly connected to the competing actions necessary to try to satisfy both an efficiency/utilization set of metrics, calculated and based on a fully burdened standard product cost, and the metrics to meet due dates for product ordered directly by the market (actual orders), demand driven flow.

If the above scenario looks and sounds familiar, it is because this same meeting takes place within supply chains around the world. The same list or some variant of this list is in every organization, even those that are considered best in class and top performers. Any organization that has two conflicting primary operating objectives and their corresponding measures, suffers from some form of the unresolved conflict. As depicted in Chap. 2, the driver of organizational waste is unresolved conflict. If most companies suffer from the above chronic and unresolved conflict, then almost every company has tremendous hidden potential.

Back to the Scientific Method

4. We evaluate the facts.
- ▲ If the facts support the explanation, we have a theory.
- ▲ If they don't, we need to go back to Step 2.

It appears the facts support the hypothesis—the plant's focus on its monthly profit through CoGS dollar credits is getting in the way of its ability to service the market. The plant is focused on a cost-centric efficiency strategy to meet its operating plan and the unit-cost targets it was built upon. The policies are designed to maximize the efficiency and utilization of all resources, minimize unit costs of parts, and maximize overhead absorption rates. The tactics applied are eroding market delivery and service, driving up the investment in inventory, increasing spending on overtime and premium freight, as well as fueling the perceived need for additional plant and equipment. In an attempt to satisfy both, the competing actions compromise the objectives the measures were intended to protect and result in lower "system" productivity, efficiency, and ROI.

Figure 5.2 summarizes the chronic unresolved conflict above, demonstrating that the conflicting actions to satisfy cost-based efficiency measures conflict with the actions necessary to protect demand driven flow. This conflict clearly shows how the unresolved conflict puts managers in a no-win set of compromises that compromises the ROI objectives. It's called a

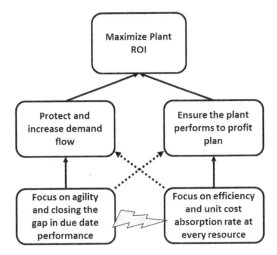

Figure 5.2 Unit Cost Centric Efficiency Strategy versus Flow Centric Efficiency
Strategy

compromise because we end up cutting short on one or both of the necessary conditions for the overall objective of the organization. This tends to
produce an oscillating effect between the two sides—a continuous set of
trade-offs. Figures 5.3 and 5.4 demonstrate the effect when the organization lives on one side of the conflict.

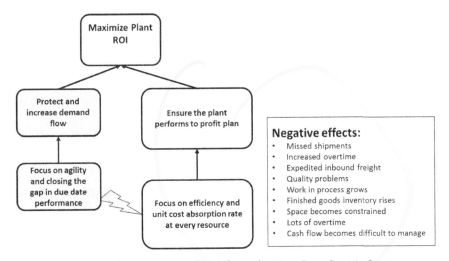

Figure 5.3 The Negatives of Satisfying the Unit Cost Centric Strategy

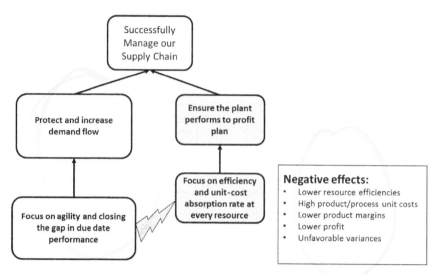

Figure 5.4 The Negatives of Satisfying the Flow Centric Strategy

The oscillation between two sets of symptoms has devastating effects on overall system performance. This was discussed in detail in Chap. 2. If the same symptoms persist, it is because we have failed to identify the root cause and end the compromise and conflict. Simply put, Company Normal created its own variation and then responded by creating even more variation. This is the very definition of management variability. They did it to themselves trying to satisfy two competing strategic objectives and the outcome is that all of the key factors of ROI go the wrong way.

ROI is a report card on the use of invested capital. It is the only real financial measure about the success of a strategy to return value to the company's shareholders. The net effect on ROI is the standard used for measuring all claims of improvement by companies in this book. If ROI is the only true financial benchmark, then it is the starting and ending point to assess the success of any strategy or tactic. A strategy that creates tactics and measures that seriously erode all of the pillars of ROI must be questioned. ROI is going down, down, down. The only way to end the conflict is to end the competition. We have to choose one primary directive to derive business rules, policies, work practices, and measures. Now we have to make a choice, a flow-centric efficiency strategy or a unit cost-centric efficiency strategy.

Unit Cost-Centric Efficiency—A Deep and Universal Truth

In 2012, an audience at an Institute of Management Accountants–sponsored workshop in Seattle, Washington, was asked to name the first law of manufacturing. The audience was made up of controllers and vice presidents of finance. Ninety percent of them worked for manufacturing and distribution firms. The audience included two companies well-known for their success and dedication to pull-based manufacturing philosophies and three university accounting professors. To a person, the answer was "cost efficiency." Three months later, the same question was posed to an audience comprised of operation and finance teams from 22 different manufacturing firms in Iowa. Without hesitation, they *all* answered "cost efficiency."

These people are well-educated, smart, and successful. They have deep experience and intuition and are rational and logical thinkers. Why they claim what they claim and believe what they believe needs to be seriously and thoughtfully explored. Assumptions need to be verbalized, verified, and/or respectfully refuted with logic and proof. That is the point of using the scientific method and is the purpose of this book. The roots of this truth are deep.

Today a cost-centric efficiency is the undisputed first law of manufacturing. It is embedded in:

▲ Policies, work practices, and measures
▲ All data compilations and/or software programs that provide management reporting visibility
  ▼ MRP planning and purchasing
  ▼ Operations scheduling
  ▼ Fully absorbed product costing and profitability analysis
  ▼ Inventory management, workforce, outsourcing, and investment decision-making
▲ All costing and financial reporting
▲ All budgeting and brand assessment
▲ All investment and labor requests
▲ Every aspect of our business, accounting, engineering and management education

It clearly drives companies to seek local improvement and stand-alone tactical initiatives at every level and resource of the business. It is insidious to the point that it is invisible. It is not questioned or thought about. All of the above clearly point out why this conflict cannot be resolved at the tactical level and any improvement gains sustained. The alignment of policies, tactics, and measures must be at a *system* and a *rules* level, which is the point of all of the conflict diagrams above. As mentioned in Chap. 4, we have a core problem at the deep truth level.

There is only one real question to answer: Are the cost-centric efficiency business rules a valid strategic driver for today's demand driven world, if the following are true?

▲ We can clearly identify the rules as a major cause of system variation and waste.

▲ We can prove the linear system assumptions the rules were built on do not apply to today's complex supply chain systems.

▲ The original basic assumptions of how costs behave underlying GAAP unit costing are flawed.

Then the time has come to consider the replacement of the deep truth of a cost-centric efficiency with the deeper truth of a flow-centric efficiency.

With the increase in complexity, our linear rules and thinking are less and less effective. The performance gap has gotten large enough to be very painful. In order to change a deep truth, being in pain is not enough. It requires understanding what to change and what to change to. If you don't know the "new rules" and how to practically apply them, then change cannot happen no matter how obvious or painful the performance gap becomes.

$$\boxed{\Delta \text{Visibility} \rightarrow \Delta \text{Variability}} \mapsto \boxed{\Delta \text{Flow} \rightarrow \Delta \text{Cash Velocity} \rightarrow \Delta \left(\frac{\text{Net Profit}}{\text{Investment}} \right) \rightarrow \Delta \text{ROI}}$$

Core Problem Area **Plossl's First Law of Manufacturing and the Demand Driven Model**

▲ Visibility is defined as relevant information for decision-making.

▲ Variation is defined as the summation of differences between what we plan to have happen and what happens.

▲ Flow is the rate in which a system converts material to product required by a customer.

We have to start with defining the rules of how costs and revenue behave because they are the key to defining what makes information relevant.

▲ GAAP and economics/management accounting have different math rules for defining how costs and revenues behave.
▲ Linear systems and nonlinear systems have different math rules for defining how costs and revenue behave.

Cost-centric efficiency rules are based on GAAP and linear system rules. Flow-centric efficiency rules are based on economics/management accounting and nonlinear rules. Before we can talk about implementing smart metrics, we need to build some common background on the rules for all of the above. The next chapters set a framework for defining and comparing the two different strategies through an exploration of the following topics:

1. Define and compare cost-centric efficiency and flow-centric efficiency as the primary directive.
   a. Compare and contrast the strategies, policies, measures, decisions, and actions.
   b. Compare and contrast their system definitions and supporting science. Efficiency follows the rules for a linear system and demand driven flow follows the rules for a complex system.
2. What is the science and mathematical foundation from which management accounting is derived?
   a. The mathematical principles in physics, economics, and statistics that make up the rules for relevant information.
   b. The old assumptions that must change because the underlying science and math models explaining complex system behavior have radically changed.
   c. The new assumptions applied to our economic and statistical models:
      i. Derive the new business rules.
      ii. Derive the associated smart metrics.
3. The history and evolution of flow and ROI as the first law of supply chain strategy:
   a. Where did they come from?
   b. Why are they the foundation of smart metrics?

4. The history and purpose behind GAAP financial accounting and full absorption costing:
   a. Mathematical principles used to derive them.
   b. Do they meet the economic test for relevant information for decision-making?
   c. Identify all of the measures/reports in the organization that use an absorption unit-cost base to derive them.
   d. Quantify the effect of the distortion collectively.
   e. Is the variation they cause large enough to justify change?
5. The case and proof that today's supply chains are complex adaptive systems, not linear systems:
   a. The differences in the rules and assumptions between linear and complex adaptive systems
   b. Translating the new rules of complex adaptive systems to supply chain management, demand driven system design, and smart metric objectives.
6. Define demand driven performance business rules and smart metrics.
7. An example of a demand driven performance system and smart metrics.

CHAPTER 6

Efficiency, Flow, and the Right Measures

The "Right" Measures

"What are the right measures?" This is a common first question. As mentioned before, the primary "right" financial measure is return on investment performance. A company cannot claim to have improved if it does not have an improvement in ROI. It is also the only measurement that makes net profit relative to the effort invested. If the net earnings are reported as $50 million for the prior period, it is only "good" in relation to the invested capital to get those earnings. It is a happy day for a small services company; not happy days at Microsoft, Boeing, or Unilever. An organization's ROI can be 4 or 24 percent; measuring it at the end of any financial period will not change the result. The tactical planning and execution decisions and the actions taken every hour, every day by people in the organization determine where the company lands on the measurement scale of ROI. Their daily use of the assets determines both the working capital position and the bulk of capital investment requests and decisions.

A secondary "right" measure for a supply chain system is due date performance (DDP). DDP is a measure of our ability to meet actual requirements, requirements that directly produce revenue. This means fill rates for make-to-stock, make-to-order, and engineer-to-order products. Again, it is the actions people take every day that determine if it is 60 or 99 percent. DDP takes a backseat to ROI because it is a component of the ROI equation. The point where the net spend, to secure an incremental improvement in DDP is greater than the cash inflow it will generate from the market, is the point where improving DDP is no longer beneficial to improving ROI. The ROI impact is the mathematical test when deciding to pursue or focus additional effort or investment to increase any key performance indicator (KPI). In fact, that is the very definition of relevant information for decision-making. ROI trumps all other measures designated as KPIs because it is *the* KPI of the total system over time. *Focus efforts where the greatest ROI opportunity exists because both time and cash are finite.*

Two critical features of both of these measures make them "right":

1. Both are measures of system output or flow over time. ROI is the flow of net cash over time and DDP is the flow of information and product that generate the net cash.

2. They are the only two measures that reflect the performance of the company as a whole system from quote to cash generation. They measure the efficiency of the entire system, thus the term *flow-centric* efficiency.

It is clear companies understand the connection between measuring product to satisfy a true market demand signal and ROI. This is evidenced by the fact that the majority of companies consider high market DDP or service levels to be one of their KPIs. It is also clear that companies understand that ROI is at the top of the food chain for measuring company performance.

ROI, however, is simply too remote to be useful for resources to make tactical decisions regarding what actions to take and in what priority to take them. In most companies, it is next to impossible for a local manager to make a connection from his or her actions today and the effect those actions will have on ROI. This has led companies to create a significant number (hundreds or more) of tactical and local measures to focus and direct people's daily actions. Most of these measures are firmly cost-centric. Let's understand why.

When a company is focused on a cost-centric efficiency strategy, the majority of these measures emphasize the area's efficiency, utilization of a cost-reduction effort based on a standard unit cost. This connection appears to make sense because there *is* a direct relationship to a decrease in *total system spending* cost and an increase in ROI. This is true as long as the *total system revenue* and *investment* remain unchanged. Cost-centric efficiency's relationship to ROI is extrapolated from a whole-system rule. The system is broken down to components, measured discretely, and then recompiled with the output of those measures. Thus the performance of the system is the sum of the local measures. *Wrong!*

Companies fail to grasp two important realities when they apply a whole-system rule to a local resource or area:

1. The extrapolation is flawed because the rules that apply to define what makes the system efficient, how to maximize the efficiency of the system, and how costs actually behave in the system, cannot be extrapolated and applied to the individual links that make up the system. A flow-centric efficiency strategy focuses on maximizing the efficiency and utilization of the entire system at the rate the market is pulling.

2. The majority of these local or individual cost-centric efficiency and utilization measures are based on the financial accounting GAAP definition of full absorption product cost. The policies, tactics, and actions to protect product gross margin and minimize product unit costs, actually conflict with the tactics and actions necessary to accomplish the other side of the ROI equation. They inhibit and block the speed of flow through the system by distorting information and directing the inappropriate use of material and resources. Thus, they erode due date performance and market opportunity. As mentioned before, GAAP was never intended to be used for planning, execution, and investment decision-making.

The starting point to understanding why the above statements are true is to define, compare, and contrast cost-centric and flow-centric strategies, their derived policies or business rules, and their tactical decisions and actions.

Figure 6.1 is a summary of a cost-centric efficiency strategy. The strategy provides the framework to derive policies, metric objectives, and tactics to deliver the strategy. Tactics have a defined system purpose, a set of actions, a schedule or time frame, and a measurable result. With tactical results, the reporting information feedback determines when, where, and how to take corrective action to protect and ensure the lowest unit product cost.

There are four basic points to understanding a cost-centric efficiency strategy:

1. The global strategy is to maximize resource efficiency and minimize product unit costs as defined by GAAP.

Alignment	Definition and Examples
Strategy	Maximize resource efficiency and utilization: **Plan and schedule resource activities to ensure the lowest product cost and highest product gross margin.** **Focus on cost reduction tactics, actions and initiatives** with emphasis on labor saving, machine utilization and inventory reductions as top priorities – **Every cost reduction increases ROI**
Policies – "The Rules"	Purchasing, Planning, Scheduling, Hiring, Pricing, Inventory management, Sales, Outsourcing, Transportation, Deployment, Engineering, Research and Development
Metric Objectives	**Gross profit product margins** – Meet the profit plan for both revenue and product cost **Part and Product Standard cost** – Efficient use of all resources **Working capital dollar targets** – Efficient use of working capital **Cost reduction initiatives** – Meet the profit plan **Product cost variance analysis** – Target resource efficiency and cost reduction opportunities/compliance
Tactics to deliver the strategy and ensure the metric objectives	Purchase least cost material, economic order quantities (EOQ) to determine production minimum batch size, extend the forecast to meet batch size minimums and minimize setups, avoid scheduled maintenance, price on volume discounts, schedule work to run on optimal (least cost) resources, invest in automation and least cost labor, restrict purchases to inventory dollar targets/turns, enforce customer order minimums, volume price discounts, ship and or buy only full trucks, outsource based on fully absorbed cost comparison savings, product gross margins used to price, promote, delete or add products. Capital investments and deletions based on fully absorbed standard cost savings.
Decisions or actions	Local resource areas take actions to maximize resource utilization and efficiency measures and minimized costs in my area of responsibility. Determine priority based on the impact to my resource area local measures.

Figure 6.1 A Cost-Centric Efficiency Strategy

2. The system is defined as linear and uses an additive mathematical equation to describe how costs and revenues behave.
3. Together, GAAP and linear system rules define/derive the policies to ensure all of the individual resource areas achieve the highest cost-centric efficiency of the resources they manage. These policies determine the tactics to plan and to execute as well as define the measurement and reporting objectives for each area.
4. The resource measurement and reporting objectives define the data collection points, the data set, and the frequency of data collection, to supply relevant information for decisions, actions, and performance evaluation.

The first question to ask is, "What determines relevance and priority in the world of cost-centric efficiency?" The answer depends on your viewpoint from inside the system. In this environment, everyone is striving to increase the efficiency of the resources they manage by driving up the utilization of their resource area. They know this behavior will drive down the unit cost assigned to each product as it passes through their resource area.

All resources, at all times, are prioritizing their actions toward this end. In an organization with a cost-centric efficiency strategy, there is very little chance of agreement on systemic priorities or relevant information to determine the system priority. It typically takes an impending crisis to get an agreement on priority. There is little to no visibility outside the local view and a limited ability to connect local actions to the effect on ROI. This environment is target-rich for conflicting measures, actions, and the resulting negative effect on ROI, such as what we see in Company Normal and the spiderweb conflict diagram (see Fig. 2.10).

The manager of a single resource does not have the same view of what is relevant as the plant manager. The engineering manager will have a different viewpoint from a staff engineer about what is relevant information. The inventory manager, the sales manager, and the purchasing manager will have a different view of what is relevant, and we can almost guarantee that their views will disagree with the controller's view. Why? Because they are viewing the system through their own local cost-centric efficiency measure rather than a viewpoint of what makes the *system* flow efficiently. They are looking through a knothole that limits their visibility to understand the system as a whole. There is, however, one assumption they will all solidly agree on: the higher they drive their local efficiency and the lower they drive their unit product cost, the more profit for the organization. It is simply the deep truth. That truth creates an obsession with measuring the individual financial aspects and implication of every resource.

Companies and managers cannot think of or even imagine another way of being. Their reporting information and their measurements are all designed to track and summarize information based on the fundamental fully absorbed unit costs inherent in GAAP. They will all agree that flow is important but cost trumps all. This is the mindset Company Normal is trapped in from the previous chapter.

Figure 6.2 is a summary of a flow-centric efficiency strategy. The strategy provides the framework to derive policies, determine the metric objectives, and the tactics to deliver the strategy. Tactics have a defined system purpose, a set of actions, a schedule or time frame, and a measurable result. The feedback of the tactical results determines when, where, and how to take corrective action to protect the tactical objectives. Real-time, visible feedback determines when, where, and who must take corrective action to protect the demand driven plan and delivery performance to the market.

Alignment	Definition and Examples
Strategy	Maximize system flow to market pull: **Synchronize demand and supply signals** between critical points ; the control and decoupling points - **Identify and Remove whatever blocks flow to and through the critical points.**
Policies – "The Rules"	Purchasing, Planning, Scheduling, Hiring, Pricing, Inventory management, Sales, Outsourcing, Transportation, Deployment, Engineering
Six Metric System Objectives	**Reliability** – Consistent execution to the plan/schedule/market expectation; **Stability** – Pass on as little variation as possible; **Speed/Velocity** – Pass the right work on as fast as possible; **System Improvement/Waste (Opportunity \$)** – Point out and prioritize lost ROI opportunities. **Strategic Contribution** –Maximize throughput dollar rate and throughput volume according to relevant factors; **Local Operating Expense** - What is the minimum spend that captures the above opportunity?
Tactics to deliver the strategy and ensure the metric objectives	Purchase as needed for quality and quantity, create no artificial batch sizes, use Constraint Throughput generation rates to determine product profitability, setups at non-constraints are free, limit the release of work to the pace and schedule priority of the system control points (constraints), invest in necessary sprint capacity to protect flow to and through the constraints, invest in strategic inventory to "position and pull" to meet customer tolerance times, offload constraint work to less efficient resources, Outsource to offload a constrained capacity resources. Use visible buffers to break variation and direct actions and priorities, Capital asset priorities based on Throughput dollar opportunity and total cycle time reduction.
Decisions or actions	Visible, real-time stock and time buffer status align priorities and identify when, who and why a decision to take a corrective action should occur. Identifying the source and impact of the variation forcing corrective action necessary to "unblock" flow prioritizes future improvement actions. Control point capacity overloads signals when to schedule overtime, push out or pull in "work" and prioritize customer promise and ship dates.

Figure 6.2 Flow-Centric Efficiency Strategy

There are three basic points to understand:

1. The global strategy is to maximize the system's flow to market pull signals. The rules derived from this strategy are used to define/derive the policies to synchronize all of the areas listed below to the same plan and execution priority.
2. Defining those policies creates the tactics to both plan and to execute as well as define the measurement and reporting objectives for each area. The focus is on creating visibility and velocity of relevant information and product flow.
3. The measurement and reporting objectives define the data collection points, the data set, and frequency of collection to supply the relevant information to satisfy the reporting and measurement objectives.

Each strategy's rules and tools define the minimum information technology each strategy needs to provide the relevant information to make decisions and judge system performance. Understanding the tactic's relevant time frame changes the definition of relevant information to enable resources to take actions to support their tactics to deliver the strategy objective.

A quick summary so far: Cost and flow have very different definitions of relevant information, business rules (policies), metric objectives, tactics, and actions:

▲ Cost-centric efficiency is focused on planning and executing the "best" individual resource efficiency and least unit-cost performance to deliver the business plan and maximize ROI. Low unit cost is a natural outcome but it has no correlation to what the system spends—the system's cost to operate in the time period measured.

▲ Flow-centric efficiency is focused on synchronizing and aligning all resource priorities to the market demand pull signal and on the velocity of system flow to maximize ROI. High DDP, short market lead times, and minimum invested capital are the natural outcomes. This has everything to do with the maximum market opportunity for the minimum system spend and investment.

These two strategies lead to different policies—business rules and conflicting and or opposite tactical decisions and execution actions. Only one of them can be optimal in terms of the system's ability to maximize ROI. One of them is based on a flawed set of rules and assumptions on how to maximize system ROI. We cannot have two opposed strategies and expect coherent planning, execution tactics, and optimal results.

Unfortunately, this is exactly where most companies and managers find themselves today—straddling a world of decisions that demand constant compromises between conflicting objectives. In Chap. 5, we used Company Normal's story to demonstrate how these measures directly result in creating roadblocks in the system's flow and constant firefighting. The result: increased spending and investment to expedite around the roadblocks, lost capacity and wasted materials, increased inventory, and ultimately lost revenue opportunity. The *total system efficiency* was decreased.

Defining Cost

Ending the competition between flow and cost metrics begins with understanding the math that each of them uses to define relevant information for good business decisions. The starting point is to agree on the definition of *relevant information* and then on the definition of *cost*. Arguably, *cost* is the most overused, misunderstood, misinterpreted, and misconstrued word in industry. Management accounting uses the economic definition of cost. It is the basis for cost-volume profit analysis. One of the foremost management accounting textbooks has the following definition to explain relevant information for decision-making:

> Every decision involves choosing from among at least two alternatives. In making a decision, the costs and benefits of one alternative must be compared to the costs and benefits of other alternatives. Costs and benefits that differ between alternatives are called relevant costs. Distinguishing between relevant and irrelevant costs and benefits is critical for two reasons. First irrelevant data can be ignored—saving decision makers tremendous amounts of time and effort. Second, bad decisions can easily result from erroneously including irrelevant costs and benefits when analyzing alternatives. To be successful in decision-making, managers must be able to tell the difference between relevant and irrelevant data in analyzing alternatives.[1]

The costs referred to in the above definition are based on how a cost behaves over a relevant range of time. In other words, time frame becomes crucial for context and relevancy.

A *variable cost* is a cost that rises and falls with the activity in volume. A variable cost is a function of the change in its cost driver, volume. This means that, once again, a change in variable cost is a function of changes in volume in the system. Variable cost is *not* tied to changes in volume of a single resource. Thus, it is important to note that while total variable cost rises or falls with volume (its cost driver), it remains constant on a per-unit-cost basis. The simplest example is raw material. This one-to-one relationship makes it easy to assign the cost to an individual product or service directly. In fact that is the definition of a *direct cost*. It is a cost that can easily and conveniently be traced to its cost object. In all of our examples, the cost object is a unit of product or a part. A line graph of a

---

[1]Noreen, Eric, PeterBrewer, and Ray Garrison, *Managerial Accounting for Managers*, McGraw-Hill Irwin, 2008, page 500.

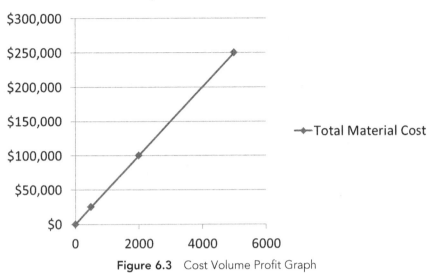

X Axis = Volume; Y Axis = Dollars

Figure 6.3 Cost Volume Profit Graph

change in variable cost, to a change in volume, will always have a slope with a one-to-one ratio of the rate of change of the unit variable cost to a single unit change in volume.

Figure 6.3 demonstrates the cost behavior and the one-to-one relationship of material cost dollars to unit volume activity. The material cost of $50 for each unit produced remains constant but the total cost rises and falls with its cost driver, volume.

A fixed cost is a cost that remains constant, in total, regardless of changes in the activity level. Unlike variable costs, fixed costs are not affected by changes in activity level over their relevant range. The relevant range is the range of activity within which the assumptions about variable and fixed costs are valid. It is also associated with time. The timeframe of the relevant range of volume is determined by the time it will take to step into or out of a fixed cost.

An example used to demonstrate the idea of a relevant range is the time frame to step into or out of a lease on a building used for warehousing. The time to step out is the life of the lease and or the length of time to sublet the space to someone else. During that timeframe, the fixed cost of the lease will not change regardless of how much of the square footage is utilized *and* regardless of what type of product is stored. The time needed to expand warehouse space, either through leasing or building, determines when the volume output limitation of space to store product is overcome. Volume cannot be increased until the space is ready to store the increased volume output. This means the volume range is fixed without an additional increase in fixed costs for the additional warehouse space. The opposite holds true as well. Reducing the volume of product stored in your leased warehouse does not reduce your fixed costs. Only a move to a warehouse with a lower cost will decrease the total fixed cost of warehousing product.

Fixed costs are easy to graph over their relevant range because they are linear and always have a slope of zero. A fixed cost is a flat horizontal line over its relevant range. What is important to understand is a *fixed cost has an inverse function directly opposite of the direct variation function of a variable cost.* As volume rises or falls, the total cost remains constant but the unit cost rises or falls with the change in volume because we are assigning or spreading the fixed cost over each unit.

Figure 6.4 graphically demonstrate the behavior of fixed costs to changes in volume activity. There is no change in total spending to either a rise in volume or a decrease in volume. The total labor and overhead for the operation remains constant with both the rise and fall of volume inside the time period of one month and a volume limit of 6000 units.

The volume limit of the current fixed cost investment is defined by the time frame necessary to step up investment to increase volume and/or

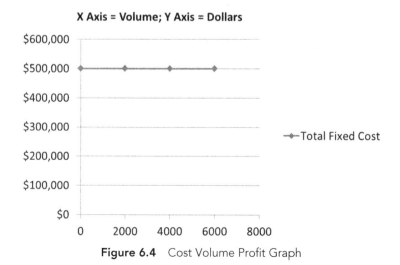

Figure 6.4   Cost Volume Profit Graph

step out of an investment and decrease both spending and volume. The relevant range defines the time frame in which the cost behavior and its volume upper limit cannot be changed. Over any one month, the plant cannot realistically add or delete headcount and/or add to capital investment to change the upper limit of volume. The total spend for all overhead remains constant regardless of a fall or rise in volume inside the 6000-unit volume limit.

Figure 6.5 is a graph of total plant spending, both variable and fixed costs, for the previously defined relevant range of 6000 units. In order to denote the relationship of total costs, the y intercept for the variable cost is not zero, but begins at the fixed cost investment of $500,000. In order to produce the first unit of product, the fixed investment must first be in place. The slope of the line is equal to the slope of the variable cost line in Fig. 6.3. A total unit cost is only true for one volume point on the graph because it includes fixed costs. It is accurate only as a representation of the past or a budget projection of a specific volume and product mix plan. The definition of the relevant range means there is no decision that can be made in the relevant range time frame that will decrease the total fixed cost spending.

Unitized fixed costs are only true for one point on the volume line. They can accurately represent past performance as depicted in GAAP

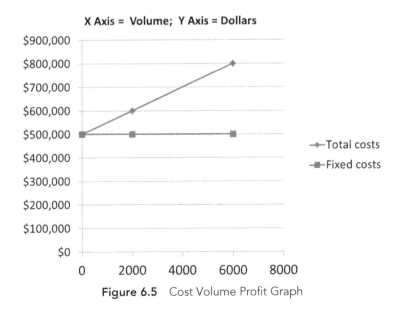

Figure 6.5 Cost Volume Profit Graph

financial statements. They are wrong at all other volume levels. They have *no* relationship to *any* cost driver inside their relevant range of time—they are fixed over the time duration and they govern the volume output.

Cost-volume-profit (CVP) analysis is a method used in managerial economics and management accounting to study the effects of output volume changes on revenue (sales), expenses (costs), and net profit (net income). In order to be relevant, CVP must make several assumptions. Running a CVP analysis involves using the differential equations above for price, variable costs, and fixed costs and plotting them on an economic graph. It can be useful for business managers making short-term economic decisions (i.e., decisions inside the relevant range of the cost assumptions). The break-even point on the graph shows the point where the sales revenue generated covers both the total fixed cost spending and the variable costs associated with the product sold. From the break-even point forward, every additional unit sold generates net profit. The difference between the selling price per unit and the variable cost per unit falls directly to net profit and is defined as the unit product contribution margin also known as throughput per unit. The mathematical formula is:

Selling price per unit – Variable costs per unit = Contribution margin per unit

CVP is a decision-making tool that expresses graphically the net profit and break-even relationship to changes in volume. CVP can be used to model the changes to break-even and net profit by varying the different assumptions regarding product selling price and product mix changes. Cost-volume-profit analysis determines the break-even point given a specific product mix, selling price per unit, and variable unit cost. Changing the product mix assumptions to understand the impact on both profit and break-even is an important use of CVP. Fixed costs are irrelevant because they do not change with a change in the inputs above. Figure 6.6 shows a CVP graph with the following inputs:

▲ Total plant spending: $50 variable unit cost, $500,000 fixed costs
▲ Volume relevant range: 6000 units
▲ Per-unit selling price: $210
▲ Contribution margin: $160
▲ Break-even volume: 3125

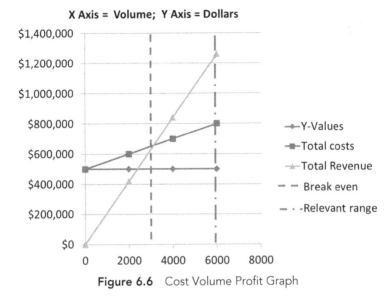

Figure 6.6 Cost Volume Profit Graph

The point where the total revenue line meets the total cost line is the break-even volume point. After the break-even volume, the contribution margin of $160 for every additional unit sold is profit. Any change in the slope of the line, such as selling price increase/decrease or variable cost increase/decrease, will shift the break-even volume point.

Figure 6.7 illustrates how unitized fixed costs behave and why they are a contrived number. No activity driver can make a unit cost valid, because it has no relationship to any volume inside the relevant range of time. A unitized fixed cost is always and only true at one point on the volume axis.

Total Volume	Unit variable cost	Total variable cost	Total fixed cost	Unit fixed cost
1	$50	$50	$500,000	$500,000
2,000	$50	$100,000	$500,000	$250
3,125	$50	$156,250	$500,000	$160
4,000	$50	$200,000	$500,000	$125
6,000	$50	$300,000	$500,000	$83

Figure 6.7 Comparison of Fixed and Variable Cost Behavior

It is 100 percent accurate to depict the past because in the past, the volume and product mix that generated the average unit costs and volume outputs are known. It is irrelevant for short-term tactical decision-making because there is no action that can be taken in the short term that will change the total fixed cost spending for the relevant time period. Regardless of the planning and execution actions taken, the minimum fixed cost spending will be approximately $500,000.

A tactical action taken for the sole purpose of minimizing a unit cost will have no positive effect on the cost side of the ROI equation. If the action impedes flow, it will most likely have a negative effect on the revenue, the operating expense cost side of the ROI equation, and a negative impact on invested capital on the other side of the ROI equation.

Management accounting textbooks are full of repeated warnings regarding the use of unitized fixed costs for decision-making. Below are several examples.

It is sometimes necessary to express fixed costs on an average per unit basis. For example we have shown how unit product costs must be computed for use in external financial statements containing both variable and fixed costs. **As a general rule, however, we caution against expressing fixed costs on an average per unit basis in internal reports** because it creates the false impression that fixed costs are like variable costs and that total fixed costs actually change as the level of activity changes. To avoid confusion in internal reporting and decision-making situations, fixed costs should be expressed in total rather than on a per-unit basis.[2]

The relevant range is the range of activity within which the assumptions about variable and fixed costs are valid. Fixed costs can create confusion if they are expressed on a per-unit basis. This is because the average fixed cost per unit increases and decreases inversely with changes in activity.[3]

---

[2]Noreen, Eric, Peter Brewer, and Ray Garrison, *Managerial Accounting for Managers*, McGraw-Hill Irwin, 2008, page 127.
[3]Ibid, page 53.

Strategic or Tactical	Plan	Schedule	Execute	Measure execution to the plan
Strategic	Annually			Annually
Strategic & Tactical	Quarterly			Quarterly
Tactical	Monthly	Monthly		Monthly
Tactical	Weekly	Weekly		Weekly
Tactical		Daily	Daily	Daily
Tactical			Hourly	Hourly

Figure 6.8 Time Horizons for Strategic and Tactical Define Different Relevant Ranges of Cost Behavior

The relevant range for all capital infrastructures in a supply chain, and in most cases this includes direct labor, is greater than the product planning, scheduling, and execution cycle time. In today's world, almost all of an organization's costs are fixed for at least a fiscal quarter (three months). If this statement is true, then there is *no* instance in which a unit product cost or a unit-cost efficiency measure is valid as relevant information for tactical planning, scheduling, and execution decisions for products or projects. Any and all unit fixed costs and or efficiency metrics that contain unitized fixed costs are irrelevant and can lead to a wrong decision.

Figure 6.8 shows the different time horizons for planning, scheduling, execution, and measurement of execution to plan. The point of the table is to demonstrate that the planning process starts out as a strategic annual financial plan that is then broken into quarterly and monthly buckets. Given a long enough time frame, all costs are variable. The relevant time frame of the decision determines the cost and revenue behavior and defines what information is relevant and what is irrelevant. Irrelevant should always be ignored. Clearly, over the typical tactical planning, scheduling, and execution supply chain *cycle time* (under three months), truly variable direct costs (materials, outsourcing costs, direct labor overtime, etc.) are the only *relevant* information for product and project planning, scheduling, and execution decision-making. At the tactical planning, scheduling, and execution level, the relevant time range is short enough that almost all fixed costs in the annual financial plan are irrelevant. They should be excluded from information for tactical planning and execution decision-making. *They should be excluded from all resource performance measures.*

Unitized fixed costs should be eliminated from all tactical planning, scheduling, and execution metrics. One of the major differences between a cost centric efficiency strategy and flow centric efficiency strategy is the cost strategy extrapolates unit-cost information from the annual to the tactical time frame to determine relevant information for planning, scheduling, and execution. A flow strategy plans, schedules, and executes to the market pull signal, using only relevant costs and revenues determined by the tactical time frame. Attempting to do both creates the dilemma of Company Normal—actions driven by irrelevant information and conflicting metrics end up compromising market delivery and financial performance.

CHAPTER 7

Our Current Accounting Measuring Mess

The obvious question is, "How did we get into this accounting measuring mess?" The short answer is twofold:

1. We have mixed up our accounting toolsets (financial versus management accounting).
2. There has been a major change in the fundamental science of systems theory, which has refuted the definition of supply chains as linear systems.

Let's start with the mix-up in accounting tools. The point of defining relevant information, relevant time frames, and the graphs depicting cost and revenue behavior is to clearly show that GAAP costing cannot be used interchangeably with *cost* as defined by economics and management accounting. They do not have the same history, rules, math models, or purpose. They are distinct bodies of knowledge and recognized as such in academia. Management accounting was created based on innovation and uses both economics and behavioral science as the foundation for its applied science. The development of the principles governing both flow and management accounting has the same history and is the subject of the next chapter. They evolved together and there is no conflict between management accounting and system flow. In fact, their development and the subsequent understanding of how to apply each of them were iterative. They secured industry dominance for their founders/inventors.

The following is a summary of the primary difference between financial accounting and managerial accounting:

> Managerial accounting is concerned with providing information to managers—that is, people inside an organization who direct and control its operations. In contrast, financial accounting is concerned with providing information to stockholders, creditors, and others who are outside an organization. Management accounting provides the essential data that are needed to run organizations. Financial accounting provides the essential data that are used by outsiders to judge a company's past financial performance. Managerial accountants prepare a variety of reports. Some reports focus on how well managers or business units have performed—comparing actual results to plan and to benchmarks. Some reports provide timely, frequent updates on key indicators such as orders received, order backlog, capacity utilization, and sales. Other analytical reports are prepared as needed to investigate specific problems such as a decline in profitability of a product line. And yet other reports analyze a developing business situation or opportunity. In contrast, financial accounting is oriented toward producing a limited set of specific prescribed annual and quarterly financial statements in accordance with generally accepted accounting principles (GAAP).[1]

Financial accounting was created to comply and be governed by a regulating body known as the Securities and Exchange Commission, the SEC. The SEC was brought into existence by legislation with the intention to protect the public and the financial market from fraudulent misrepresentation, by companies, regarding the reporting of annual and quarterly financial statements. The following is from the official website of the Financial Accounting Standards Board (FASB), an SEC organization.

> The SEC has statutory authority to establish financial accounting and reporting standards for publicly held companies under the Securities Exchange Act of 1934. Throughout its history, however, the

---

[1] Ibid, page 2.

Commission's policy has been to rely on the private sector for this function to the extent that the private sector demonstrates ability to fulfill the responsibility in the public interest.

Since 1973, the Financial Accounting Standards Board (FASB) has been the designated organization in the private sector for establishing standards of financial accounting that govern the preparation of financial reports by nongovernmental entities. Those standards are officially recognized as authoritative by the Securities and Exchange Commission (SEC) (Financial Reporting Release No. 1, Section 101, and reaffirmed in its April 2003 Policy Statement) and the American Institute of Certified Public Accountants (Rule 203, Rules of Professional Conduct, as amended May 1973 and May 1979). Such standards are important to the efficient functioning of the economy because decisions about the allocation of resources rely heavily on credible, concise, and understandable financial information.[2]

At the time of this writing, the seven-member governing board of the FASB is made up of one academic, a former industry CFO with a previous career in public accounting, two Wall Street firm executives with previous careers in public accounting, two former partners in a public accounting firm, and a former forensic auditor whose career path was also through public accounting. The point is, they are experts in GAAP and auditing standards.

GAAP's rules are built on an amalgam of legislation to protect the public, industry lobbying to protect the interests of the various industries, and promulgated by a committee process. The importance of flow and consistent and appropriate application of mathematical principles to manage manufacturing and supply chain operations has nothing to do with its purpose. Its rules are about the classification, categorization, and grouping of data points to summarize the past activity and present an accurate static picture of value on a specific date in the past. The purpose is to ensure that the assets of all companies, in a common industry, present their information using the same rules. The objective is to ensure fair and consistent financial statement comparisons between time periods for a company and

---

[2]http://www.fasb.org/facts/

between companies. Audited financial statements are said to be fairly presented if they follow the GAAP rules when they are compiled.

The purpose of financial audits is to verify and certify that the financial statement presentation is in accordance with GAAP rules. The mathematical rules they use to compile are additive and are only intended to model the real cost and revenue behavior of the company for a past time period. They *do not* infer any statement about future performance or potential, and they are not intended to be used to provide relevant information for future decision-making. Audited financial statements based on GAAP are not limited to the United States; they are now a worldwide reporting requirement and are governed by a variety of different agencies or entities. For example, "after the fall of Communism, accounting methods were changed in Russia to bring them into closer agreement with accounting methods used in the West. One result was the adoption of absorption costing."[3]

Financial accounting has a valid purpose and it is not our intention to trivialize the AICPA, the FASB, or the SEC. They are necessary and serve a very important purpose for shareholders and financial markets. Our purpose is to simply point out that you will not get the result you want if you are using GAAP rules, reporting, and measures to make decisions in manufacturing and supply chain operations. GAAP math simply does not reflect the mathematical principles of the sciences for managing dependent event economic systems. The entire purpose of financial accounting and fully absorbed product cost accounting is to account for costs in the past and fairly present a snapshot in time of a company's past performance. You can only manage people and resources in the present with information that is relevant to predict the outcome over the future time frame relevant to the decision.

Remember, the entire purpose of cost accounting and financial accounting is to report a company's past performance in a manner consistent with generally accepted accounting principles—GAAP. It exists to provide a set of common rules to make the *past* performance of one company comparable to the *past* performance of another company, or one *past* time period for the same company comparable to another *past* time period of the company. Full absorption accounting requires that all manufacturing costs be assigned to

---

[3]Enthoven, Adolf J.H., Russia's Accounting Moves West, *Strategic Finance,* July 1999, pp. 32–37.

a product to produce a fully absorbed unit cost for the purpose of valuing inventory and the cost of goods sold in the income statement.

GAAP compliance (or its regional equivalent) is a fundamental requirement for doing business in much of the world, but you are not required to use it for decision-making. In fact, management accounting does not recommend the use of full absorption costing for decision-making; a material distortion is almost guaranteed to occur.

It is important to understand how the assumptions differ between the two sets of accounting as well as why they have gotten mixed up. You will get a predictable result when GAAP unitized-cost information is used to make decisions, as demonstrated with Company Normal in Chap. 2. However, it is not the result desired or intended by the company's management. The gap between what is relevant information and what is being provided to business decision makers as relevant information is growing.

The confusion and the gap between the two began to be noticeable in the 1970s. It has grown increasingly more prevalent and acute. With a solid understanding of how costs really behave—the economic definition of variable and fixed cost behavior in Chap. 6, it is clear fixed costs do not behave as they are accounted for in full absorption GAAP financial accounting. This isn't new news. Numerous business experts and academics inside and outside of accounting have been researching and writing about this subject and the negative results of using GAAP financial costing since the late 1980s. What is new is the recognition that there is more to the core problem area. The solution is not simply to turn back the clock to the 1960s from an accounting perspective.

The gap between the two accounting perspectives has become more prevalent for three reasons. The reason for its acuteness will be addressed separately later in this chapter.

Reason 1: The Proliferation of ERP/MRP II

The first reason is that more and more manufacturing and distribution companies have purchased and implemented ERP and MRP II information-technology platforms. In 1975, only 700 companies had an MRP system.[4]

---

[4]Orlicky, Joseph, *Material Requirements Planning, The New Way of Life in Production and Inventory Management*, New York: McGraw-Hill, 1975, Preface, page ix.

Today, it is common to find ERP (with MRP II embedded in it) packages even in very small manufacturing and distribution companies.

The adoption of materials requirements planning (MRP) began in the 1970s. By the early 1980s, companies using MRP II had embraced using automated costing roll-up structures to speed closing their month-end financial statements. One of the promises of the technology was the elimination of middle management positions to manually compile information. A standard cost roll-up system was fast, accurate (no computational errors), and could speed up month-end closing as well as eliminate accounting analyst positions. The routing, part, and product structure records were all formatted to accommodate automated absorption costing. Manufacturing systems that were originally designed to capture standard routing time and usage inputs (units of measure and the cost of the material input) for manufacturing management were now primarily focused on being a costing system for GAAP.

The purpose of cost roll-up is to include the cost of items purchased and manufactured at lower levels (materials and intermediate manufactured and assembled items) in the costs of the material located at the top of the structure—end item part. Every lower layer cost is rolled up to one number on the next level of the bill of materials structure. The one number reflects the *fully absorbed cost* of the lower level components as if it were a purchased component. The cost components include the costs of material, labor, overhead, outside services, freight, duties, and setup. However, today some ERP systems for multilevel production structures also feature the ability to contain costs that should not be rolled up for GAAP; cost of goods sold. This added functionality is the result of activity based costing advocates and includes such costs as engineering, research and development, sales, and administration costs. In customizing for product-cost planning, you can define for each cost component whether the assigned costing results should be rolled up or not. This actually exacerbates the potential for distortion.

Prior to MRP II, relevant variable cost information was available. As an example, in the early eighties, Debra Smith (co-author of this book) was the director of finance for a division of the Clorox Company. They had three distinct costing functions. The first was a cost accounting manager, whose job was to make certain the company captured the necessary product cost information to support the external financial reporting requirements. They reconciled standard cost variance accounts and interacted with the public auditors. The second was a general accounting manager who took care of the general ledger, accounts payable, accounts receivable, and also interacted with the public audit firm. The third was the manager of analysis and control. He was responsible for providing relevant information to operations, sales, and marketing. All of his analyses were based on the contribution margin of each product category (selling price less truly variable cost, excluding labor), also known as through put dollars. It was his job to explain the variances between the actual and planned net income and cash flow, based on differences in actual to plan for volume, selling price (promotion and full price mix), and variable cost for each individual product as well as the variance in product mix. Marketing and Sales used this information to spot product trends, plan promotions, and communicate with both the sales force and distributors. All fixed cost variances from budget to actual were evaluated as line items. This included direct labor, as the plants were highly automated. No unitized fixed costs were used to make any internal marketing or investment decisions or considered in plant scheduling, planning, and execution measures.

Where have all of the management accountants gone? They have become cost accountants or have been squeezed out altogether. Breakthroughs in production technology have greatly reduced the direct labor content in manufacturing. Indirect labor and capital investment (overhead) have replaced direct labor inputs—someone has to feed and care for the technology. The focus has shifted to indirect labor for cost-cutting opportunities. ERP and MRP II systems were sold partly on their ability to automate cost roll-ups and eliminate the accounting tasks of compiling data, speed reporting, and reduce indirect labor. The irony, of course, is that IT departments (indirect labor) are as big as ever.

There is a reason why every new ERP implementation begins with the financial module and the costing system. They cannot turn off the old system until they can generate financial statements in the new system. Remember, GAAP requires consistency for fair presentation. Thus, the old bad habits become the new bad habits. Most companies compromise on the set-up of their manufacturing system in favor of speed to get the cost system up for financial statement generation. Accuracy of routings and standard time and material inputs as well as choices on the output are often compromised, rendering the manufacturing system unsatisfactory for the purpose it was designed for—planning, purchasing, and scheduling. Worse yet, the "cost information" it provides is *irrelevant* for planning and execution decisions.

Including irrelevant information in decision making can lead to a sub-optimal and in many cases even a detrimental decision/action. The graph in Chap. 1 (Fig. 1.1) verifies the scope of companies' dependency on the use of adhoc systems (primarily spreadsheets) to augment planning.

The problem is more than a disagreement on the strategy, policies, measures, and execution actions to maximize ROI. Even if a company wishes to make the change, there is a fundamental lack in companies' ability to extract the needed relevant information from the oceans of irrelevant but "accurate" data. The way their ERP/MRP software has been programmed to collect and roll cost data or the way companies defined and implemented their MRP II costing module, simply preclude access to the relevant information.

Reason 2: The Growing Distance between the Front Office and Operations

The trend in the last 15 years is the leveraged consolidation of many different manufacturing entities: conglomeration. The result is a top management structure increasingly removed from the realities of manufacturing flow and supply chain logistics. With less intuition, the more reliant management has become on the financial reporting information generated by their ERP systems. "What do the numbers say?" The computer-generated cost and variance analysis reporting is perceived as the only reliable and accurate source of information to understand, analyze performance, and guide monthly course corrections. It is understandable, as these executives must answer to "The Street," or at a minimum be able to address ownership's assessment of their financial performance to stated plan (usually in quarterly buckets). This puts increased emphasis and management focus on working capital ratios as well as pressure to meet both their quarter and annual revenue and profit plan.

Reason 3: The Fading Away of Management Accounting

The trend to de-emphasize management accounting by both business and academia and conversely emphasize *new financial costing* techniques has greatly added to the confusion. This trend has increased, regardless of the dubious mathematical validity to provide and meet the test for relevant information. Any method that attempts to assign fixed costs regardless of the cost driver is simply arbitrary and will often result in dubious information and dubious conclusions. It is complex, messy, and violates mathematical the laws of how cost and revenues behave.

The trend of blurring management accounting and cost accounting became apparent in the early and mid-1990s and it appears to be increasing. For a variety of reasons, management accounting has come to be perceived as old fashioned and out of date. This is a true statement but for all of the wrong reasons. Management accounting needs to be updated, along with all the rest of the applied sciences (but more on that in the next section). Providing relevant information is even more critical today because of the speed and quantity of information that is available. Management accounting is relevant and has a powerful role to play in a demand driven strategy and smart metrics.

The trend of de-emphasizing the key management accounting concepts in favor of cost accounting and financial statement analysis contributes to the misunderstanding on the very different roles and uses of financial cost accounting and management accounting. Despite management accounting textbooks' repeated warnings against the use of full absorption cost accounting and variance analysis for decision-making, as well as the pitfalls of relying on allocated fixed costs for product decisions, companies are stuck in it. In business schools, management accounting required coursework has shrunk to one required quarter, and/or has been entirely merged and replaced with textbooks entitled "cost management." The authors find this term and title disturbing on two levels.

First, the title "cost management" misleads students into believing "costs" can actually be managed and the arbitrary "drivers" chosen to unitize fixed costs can be managed by an action at a local resource or discrete process level. Managers make capital investment decisions and once they are made, the costs are sunk; they cannot be undone by a decision on how to utilize them. Managers manage people and resources, *not* "costs"—costs are accounted for. That is the whole point of financial accounting. Money is either spent on something or not. It is then classified to the correct general ledger account. It has either happened or not. You cannot manage a past event, no matter how much you may regret it.

Second, the trend toward combining the curriculum of management accounting and cost accounting has blurred the lines between two fundamentally different bodies of knowledge with very different purposes and information users. There is no field or body of knowledge known as "cost management." Presenting it as a smorgasbord of costing options that are relative choices rather than different tools for different purposes does not make it one.

The gap in relevant information has become more acute for two reasons.

Gap Explanation 1: An Increase in Complexity

As demonstrated in Chap. 1, the dramatic increase in the complexity of modern supply chains has exacerbated the shortcomings in the current state of the technology and application of today's materials and distribution management and planning tools. A summary of the problem and the solution was explained in Chaps. 1 and 2. An in-depth understanding of the problem and the solution, demand driven MRP (DDMRP), is contained in the third edition of *Orlicky's Material Requirements Planning* (Ptak and Smith, McGraw-Hill, 2011). Management accounting, like MRP, is even more critical in today's complex supply chains. It needs a similar rules update, as seen with DDMRP.

Gap Explanation 2: The Shift from Linear to Nonlinear Systems

Breakthroughs in the hard sciences have resulted in a fundamental change in how science understands, explains, and models the laws governing how complex systems behave. As supply chains have become increasingly complex, the gap between the current thinking and tools to manage linear systems and the need for more robust approaches is becoming increasingly obvious. The thinking and mathematics of this profound change is only beginning to be recognized as both valid and necessary to the applied sciences. Economics, operations management, accounting, and all other business curriculum and the current rules they are built on, are still in the proverbial Dark Ages when it comes to understanding and addressing complex systems. In short, applied sciences foundational rules have changed but these applied sciences have not.

A serious gap exists between what is now understood and accepted as the rules governing complex systems and the rules currently being taught and applied to manage supply chains. As mentioned previously, the cost and revenue behavior rules currently being applied in economics and statistics and as taught in the current coursework in business schools are inadequate and fundamentally flawed. This is understandably a tough pill to swallow for everyone invested in the status quo. We should remember that the hard sciences struggled for over half a century to comprehend and integrate the shift from unexplained anomalies and unexplained cracks in Newtonian physics to mainstream acceptance of quantum mechanics, chaos theory, and complexity theory. The gap finally became just too big to pretend it did not exist and mainstream science has made the shift. This shift is no less dramatic than was experienced during the Industrial Revolution with the breakthroughs generated from Newtonian science. The productivity gains possible today are equally as staggering for those who are willing to embrace the science and rules governing complex systems and flow.

Only recently (post-2000) is there serious business research and writing on complexity systems theory. One example of this new focus on complex adaptive systems is a 2008 paper, published by the *International Journal of Physical Distribution & Logistics Management*. The authors, Wycisk, Mckelvey, and Hu Eismann, cite more than 79 references in support of their paper. Their abstract, as well as their opening sentence, confirms this shift is already here. Supply chains are complex systems and it is increasingly being accepted that they are in fact complex adaptive systems.

There are three main points to their article that every business leader, manager, and academic must understand because they are game changers, shifting the very foundation of the applied sciences:

1. The understanding of supply chains as logistics systems has evolved over time from linear structures to complex systems.
2. Complexity science uses Pareto distributions to explain well-known phenomena of extreme events in logistics, such as the bullwhip effect.
3. The evidence is growing that supply networks are complex adaptive systems (CAS) and that changes all of the assumptions about how best to manage them.

Complex adaptive system rules are the science foundation for demand driven strategy and smart metrics. This will be explored in-depth in Chaps. 10 and 11.

Silo Thinking

We live in an interdependent world. Science has made and is making breakthroughs in understanding complexity's place and in explaining the underlying order of systems. Artificial boundaries have been created when we study our world, and this is just as true when we approach managing a supply chain. The separation makes it easier to study or teach, manage, or measure a subject, but we lose the fundamental understanding of how one kind of knowledge relates to other kinds of knowledge. In addition to losing the inherent interconnectedness, distortion occurs in the translation from a hard science to a practical application—the applied sciences.

The applied sciences—logistics, engineering, accounting, marketing, and business operations management—"dumb it down" for practical application. Practical, defined as good enough to accomplish the goal, works *until* a major breakthrough occurs and changes the underlying principles and mathematics of the hard sciences. This is a gap all of the applied sciences are currently experiencing. The train has left the station and the applied sciences are not on it! In fact, supply chain management is currently on a different train, heading the wrong way with GAAP and its misapplied math at the throttle. In the world of science, the keystone is mathematics. One of the world's first scientists, Galileo Galilei, described it well 500 years ago.

> The universe is a grand book, which stands continually open to our gaze, but cannot be understood unless one first learns to comprehend the language and interpret the characters in which it is written. It is written in the language of mathematics.[5]

The hierarchy of scientific disciplines is illustrated in Fig. 7.1. At the bottom of the diagram is the source and the foundation to everything stacked above. Mathematical knowledge is the tool that allows us to describe what happens in each successive field of knowledge as we move up the figure. The mathematical principles used to derive the applied sciences are the basis for the confidence in its rigor and validity. The adherences to the mathematical underpinnings of the foundation determine the fitness of the tool for its purpose. Every level of the pyramid must be built in harmony with each science below it in order to remain truly scientific and its assumptions and conclusions valid. *A fundamental change in a foundation science translates to the need for a fundamental change above.* Figure 7.1

---

[5]Machamer, Peter, The Cambridge Companion to Galileo, (Cambridge, U.K.; Cambridge University Press, 1998): page 64.

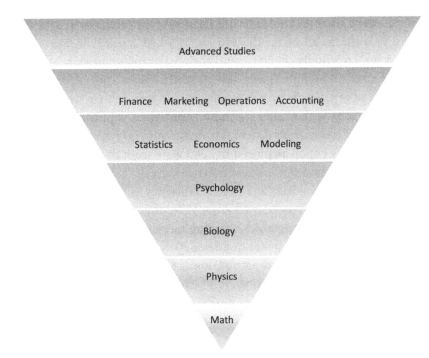

Figure 7.1   The Hierarchy of Scientific Disciplines

makes perfect sense until you get to financial accounting. The way financial accounting principles were and still are derived has no foundation in science. Its rules for application and mathematics are imposed regulations, legislated by Congress, overseen by the SEC, and written by a committee of individuals from public accounting and investment banking.

All of the applied sciences are currently struggling to understand the changes in the foundation sciences and their impact. The basic concepts of economics and management accounting are still valid today but they need to catch a ride into the twenty-first century. They are still at the train station waiting for the new train!

Challenging a Deep Truth by Definition Shakes Up the Status Quo

The discovery of complexity science changed the very foundation of mathematics and the hard sciences; it requires every level of the knowledge pyramid above them to consider and change accordingly, or risk irrelevance. This means all of the derived and applied sciences in universities must fundamentally challenge and change the rules and principles to include complexity and nonlinear systems mathematics. Business owners, managers, and organizations must challenge and change their thinking in order to change how they manage and measure their supply chains. The supply chains of today are as different from the supply chains managed in 1970 as the supply chains managed in 1860 were from the supply chains of the 1920s when modern business practice came into being.

The best way to validate flow-centric efficiency as the system strategy and ROI as the primary system measure is to understand its roots, history, and evolution. This requires breaking the history of the connection between science and business into two distinct eras. The first period birthed and emphasized management accounting (1890–1980), and the second emphasized GAAP cost accounting (1980–2013) for relevant information. Chapter 8 will discuss both of these periods in detail.

The Evolution of Flow and ROI as Strategy

The first 30 years of the twentieth century saw a fundamental shift in business equal to the dramatic evolution supply chains have experienced in the last three decades. The laws of physics, statistics, and mathematics as defined by Newtonian theory created huge breakthroughs in technology, business productivity, and transportation. They necessitated the invention of new management methods and measures that are still being applied today. Both flow and ROI were outcomes derived and perfected during this period of business innovation. Relevant information for decision-making was derived from necessity—the mother of all inventions. The important thing to understand is that the evolution of management accounting was always the result of an opportunity to obtain a competitive edge.

Management accounting evolved from a series of innovations driven by the need to control and measure evolving forms of business unseen prior to the late 1800s. These new forms of business were made possible by the breakthrough in technology and automation using Newtonian science. Management accounting was not derived from economic theory, but rather was the evolutionary brainchild of a series of pioneering individuals who, out of necessity, created measures and reporting to understand how costs and revenues behaved, to manage people and resources, and to make capital investment and policy decisions. In every instance, there was a competitive need they wanted to understand and overcome. Their methods proved themselves in the capital arena by propelling the users to the top of their industries. Each of these pioneers had to break out of the thinking of their time in order to create a solution that pushed through the limitations that limited previous business models.

The innovations of these industrial giants paralleled the innovations of Frederick Winslow Taylor, who first applied the Newtonian methodology of science to business management. He was the father of scientific management and arguably one of the most misrepresented and misunderstood individuals in history. His impact on manufacturing is still seen today. Taylor understood that the scarcity of skilled labor constrained the rate of output of all industries of his time, and he set out to break that constraint.

The paths of these giants, their ideas, and methods collided in the first three decades of the 1900s and became the underpinnings of modern management accounting, planning, scheduling, industrial engineering, and logistics. This collision took place in the fledgling automobile industry where the giant of physical flow, Henry Ford, met the giant of relevant information, F. Donaldson Brown. Their face-off would alter American industry forever.

Henry Ford understood how to break and exploit physical constraints that blocked production flow. He understood that the efficiency of the system was the only efficiency that mattered. He systematically identified physical and thinking constraints that blocked the rate of physical flow and removed them. Donaldson Brown understood the overall goal of every business should be to use its assets as effectively as possible to make a profit. He invented the DuPont ROI equation and financial ratio analysis, the first combination of cost management with asset management. He used cost-volume-profit analysis to segment the automobile market with innovative financing and market segmentation to compete against Ford. His ROI equation and ratio analysis are still the standard today.

The Birth of Product Costing

Prior to 1810, businesses were small, employed few people, and had little overhead investment. The market price of the materials purchased added to the time to assemble a simple product was good enough to make pricing decisions for a product. The inputs equaled the outputs and everything necessary to make a decision was visible. Supply chains tended to be groups of family-owned businesses in close physical proximity to each other. They each focused on doing one thing well. Small family businesses were loosely grouped and largely independent of each other.

By 1814, the Industrial Revolution that had begun in Britain had migrated to the east coast of the United States. Mass production of a single good became the norm. The inputs still matched the outputs. The textile industry is an often-used example. Instead of using small family-owned businesses referred to as "cottage industry" to card, spin, and sew material into cloth, raw cotton came into one end of a factory and a group of workers, organized by specialty, turned raw cotton into cloth. These circumstances were the first step in the birth of management accounting. History shows that these early textile mills developed a remarkably good accounting system and separately tracked direct and indirect costs of manufacturing.

A direct product cost (a cost directly and easily traced to a product) was separately recorded from indirect costs (everything else). Workers were paid per piece. An inefficient worker was paid less than an efficient worker. No output equaled zero direct labor cost. In order to price a bolt of cloth, the cost had to be understood from the cost of the delivered raw cotton to the delivered bolt of cloth. They carefully noted the efficiency of cotton used, labor time, and general overhead. Product costing and efficiency as applied to labor and raw material usage was born. Variable costs were assigned to products, and fixed costs were assigned to the period they occurred (monthly) for monthly profit-and-loss statements.

Sales – Variable product costs
= Contribution margin – All overhead spending
= Net profit

The Birth of Decentralized Management

In 1869, Leland Stanford drove the ceremonial last spike that linked the Pacific to the Atlantic, and railroad companies became even bigger giants of industry. They also presented a hierarchical challenge never seen before. Not only were managers managing managers, but the geographical spread was enormous, the volume of transactions high, and communication technology relatively primitive. A senior vice president of the Louisville & Nashville Railroad, Albert Fink, solved the problem by tracking operating expenses with a calculation called *cost per ton-mile* (the average cost to move a ton of material one mile). In addition, organizations used a ratio of operating expenses to revenues to evaluate individual managers' performance as well as understand each area's impact on the total financial performance. Variances from the standard cost and revenue were used to pinpoint cost differences and investigate the cause.[1]

An Accelerant—Andrew Carnegie

Andrew Carnegie began in the textile industry as a bobbin boy, moved to a management position in railroading, and then brought all of the lessons he learned in both industries to the production of steel. He elevated the study of cost and output per unit of cost to an entirely new level. Every department at Carnegie Steel reported on the amount and cost of materials and labor on each order of steel as it passed through their production zone. The idea of *job costing* was born. Carnegie personally queried his department heads for the reasons for cost changes. Charts of costs were reviewed to check the quality and mix of raw materials, evaluate improvements in processes and products, and to price contracts. Carnegie's practice codified material variance analysis and established it as common business management practice.

> At this point in time, management accounting, known as cost accounting, was strictly focused on cost measurement.[2]

In the nineteenth century, the steel industry was controlled by skilled workers who decided how work was to be done. There were few if any foremen. Carnegie was possessed by technology and efficiency in a way no businessman before him had ever been. He plowed profits back into his capital structure automating processes, breaking the constraint of skilled labor, greatly increasing the productivity per man-hour, and decreasing the number of laborers and their necessary skill sets. The transfer of knowledge from skilled labor to machines accelerated. His relentless efforts to drive down costs and undersell the competition made his steel mills the most modern in the world, models for the entire industry. Carnegie used these gains in productivity and his cost accounting methodology to direct capital improvements, make productivity gains, lower prices, and drive out or acquire his competitors.

---

[1]http://blog.richmond.edu/acct305s11/files/2011/01/history-of-mgmt-accounting.pdf
[2]Ibid.

Borrowing from his experiences in textiles and his observations of Britain's steel industry, he invested heavily in vertical integration. Carnegie and his partner Henry Clay Frick controlled vast holdings of coal mines, furnaces, and mills, along with railroads and ships to connect them. Frick and Carnegie understood the importance of system flow and efficiency. They sought to vertically link together their supply chain both to control supply and speed flow in order to keep prices down and drive consumption.

Carnegie described himself as a steel man and only a steel man. Despite their massive size and technology (comparative to the time) steel had very few products and product variants; it was truly a commodity market. Carnegie relied totally on a key index ratio *tons per hour* to measure productivity gains and price steel. It became the industry standard and is still the standard today for steel. Carnegie's concern was almost wholly with prime costs (direct costs of labor and material). He and his associates appear to have paid almost no attention to overhead and depreciation. Carnegie relied on replacement accounting by charging repair, maintenance, and renewals to operating costs. Carnegie had, therefore, no certain way of determining the capital invested in his plant and equipment. As on the railroads, he evaluated performance in terms of the operating ratio (the cost of operations as a percent of sales) and profits in terms of a percentage of book value of stock issued.[3]

It is safe to say that some version of costed output per hour is a prime measure in almost every manufacturing industry. However, the measure and how it is applied have very different consequences in the modern world of complex supply chains for three reasons, all of which have been explained in previous chapters:

1. Global surplus capacity
2. The proliferation of different products and SKU associated with them
3. The inclusion of fixed costs in the cost output per hour computation

The important thing to note is the value of how well it worked for Carnegie to measure productivity gains from invested capital or process changes as well as a method for focusing when a process deviated from the standard time and standard chemical "recipe" and material input to output.

---

[3]Chandler Jr., Alfred D., *The Visible Hand: The Managerial Revolution in American Business,* Harvard University Press, 1977, page 268.

These are still a valid use of variance analysis. In Carnegie's day, there were no attempts to assign overhead or account for overhead as a unitized cost per ton, and steel was a commodity with very few product SKUs.

The Rise of Wholesale Distribution and Large-Scale Retail

"The ability to move goods by rail enabled large companies to develop and produce extremely high volumes of goods for consumption by the American public. This led to the emergence of a new breed of business—wholesalers and retailers. Besides making many diverse items available for purchase from one source, these wholesalers and retailers provided other critical services, including distribution and delivery and credit service on account."[4]

Managers and accountants in this industry learned to focus on a very important idea: move the inventory!

The success of the mass merchant hinged on inventory turnover, called *stock-turns*. By selling goods faster than smaller local merchants, large-scale wholesalers and retailers could charge lower prices and still realize tremendous profit. Up to this point, big business in America had focused almost exclusively on cost. At the end of the nineteenth century, wholesalers and retailers introduced a new concept to management accounting. Companies could make a lot of money by controlling and evaluating the way managers use *assets* (in this case, inventory). As early as 1870, Marshall Field and other large-scale retailers began monitoring stock-turns throughout their organizations with great interest.[5]

This was an important step towards the modern-day techniques of asset (or capital) management commonly used today in manufacturing, distribution, and retail. These large-scale retailers understood the rate the organization made money was at the rate of contribution margin turns per unit of square footage. Departments won and lost space based on their rate of turn per use of their constraint, which was display space.

---

[4]http://blog.richmond.edu/acct305s11/files/2011/01/history-of-mgmt-accounting.pdf
[5]Ibid.

The Birth of Conglomerates and Management Accounting

This brings us to the DuPont family. They were the first to combine many different types of businesses under one corporate umbrella and capital management structure, geographically dispersed across the United States. Each of the businesses required monitoring, attention, and capital if they were to grow and prosper. They quickly realized the scarce resource they had was time and capital. They needed to understand how to make trade-off decisions on where to invest their time and capital. Previously all of the ratios on efficiency and stock turns had been used to compare like operations. They were virtually meaningless, however, as a comparison between such different business entities.

Enter the father of modern management accounting and relevant decision-making, F. Donaldson Brown, an electrical engineer turned accountant (yes, you read that correctly). He recognized that the overall goal of every business should be to use its assets as effectively as possible to make money.

Comparing net profit alone was insufficient; there must be a ratio of net earnings against the capital invested. Although the idea of return on investment (ROI) was not new at the beginning of the twentieth century, Brown turned ROI into a comprehensive performance-measurement system. He was the first to point out that if prices remained the same, rate of return on invested capital increased as volume rose and decreased as it fell. He invented cost-volume-profit analysis.

These tools allowed the DuPont cousins to manage to outstanding success. Pierre DuPont and F. Donaldson Brown took this technique with them in 1920 when they went to rescue General Motors, a struggling automobile manufacturer. The DuPont ROI formula shown in Fig. 8.1 revolutionized business management. It is a cornerstone of management accounting and is absolutely relevant today.[6]

If Fig. 8.1 looks familiar, it should. It is widely used and is as relevant today as when Brown used it. The formula as developed by Brown and used successfully in Du Pont and later General Motors, consisted of a Cost of Good Sold (COGS) formula with only variable costs—no unitized fixed costs. These giants all understood that the rate of flow directly determined

---

[6]Davis, T.C., How the Du Pont Organization Appraises Its Performance, in *AMA Financial Management* Series No. 94. (American Management Association, New York, 1950), page 7.

Figure 8.1 The DuPont ROI Formula Developed by F. Donaldson Brown

their profits and growth. The efficiency of that flow was governed by the rate, speed, and use of the inputs to satisfy market demand. *In other words, they knew flow and efficiency are not in conflict if the focus is maximizing efficiency of the entire system as measured by rate of net cash generation— ROI.* The conflict between absorption costing and variable costing did not exist. GAAP accounting and its incursion into operations decisions had not yet come into existence.

Table 8.1 is a summary of the evolution of management accounting, the innovation created, and the competitive advantage it provided.

Automation and the Death of the Craftsman

The evolution of the productive gains from breaking tasks apart, knowledge transfer from people to machines, and the separation of planning and scheduling from execution were accomplished in the first 25 years of the twentieth century. The world changed dramatically and required entirely new management methods.

No one was more influential in driving Newtonian science into business than the father of scientific management, Frederick Winslow Taylor. His innovations in industrial engineering, particularly in time and motion studies, paid off in dramatic improvements in productivity gains for companies that embraced his methods. It is not an understatement to

Table 8.1 History of Management Accounting

Development	Innovation	Competitive Advantage
Textiles industry, 1814	Product costing and efficiency as applied to labor and raw material usage. Variable costs were assigned to products and fixed costs were assigned to the period they occurred (monthly) for monthly profit-and-loss statements.	A reporting system to plan, control, and evaluate work being done by others in a factory setting.
Railroads, 1860s	Development of operating ratio of expenses to revenues to provide competitive information to judge performance of managers in a hierarchical structure.	Performance measures can be used to delegate responsibilities and to control and evaluate the business from a distance and allow companies to spread out geographically.
Steel, 1890s	The idea of job costing and variance analysis was born. The amount and cost of materials and labor on each order of steel was recorded as it passed through each production zone.	Carnegie used gains in productivity and his cost accounting methodology to direct capital improvements, make productivity gains, lower prices, and drive out or acquire his competitors.
Steel, 1890s	Charts of costs were reviewed to check the quality and mix of raw materials used to evaluate improvements in processes, products, and to price contracts.	Cost accounting moved out of the back room to the manufacturing floor and was used to evaluate input differences to standards to ensure both process control and direct improvement.
Wholesalers and retailers, 1870s	The success of mass merchandising depended on inventory turnover or "stock turns."	The first focus on making money by controlling and evaluating the way managers used assets.

(Continued)

Table 8.1 History of Management Accounting (*Continued*)

Development	Innovation	Competitive Advantage
DuPont, 1903	The DuPont ROI formula in Fig. 8.1 provided the framework to understand how to make trade-off decisions on where to invest time and capital across different types of businesses, geographically dispersed. Variable budgeting, cost-volume-profit analysis, and asset ratios. Combined and consolidated the three basic types of accounting: financial, capital, and cost.	A comprehensive performance management system designed to direct the most effective use of assets to make the most profit and measure the results. The ability to make trade-off decisions to understand the profit implications of pricing, variable costs, and investment decisions. Variable budgeting to control inputs to match production were controlled and tied to the anticipated market demand. Prices, unit costs, and rates of return were all closely related to the volume permitted by demand.

say they have been universally embraced. The major elements of Taylor's scientific management are recognizable today. He would probably be shocked to see modern factories and the extent to which work has been both standardized and automated.

Taylor referred to the elements of the scientific management approach as "the elements of details of the mechanisms of management." Taylor developed and advocated:

1. Time standards for every task
2. Standardization of tools
3. Standardization of work methods
4. Standard work instructions
5. A modern product costing system—variable costs or production
6. Separate planning function—implement a standard product routing system
7. Management by exception principle (variance analysis—focus on "out of standard" areas to improve)

8. Large bonuses for successful performance based on output rate to the standard time

9. Pay for output produced with an emphasis that labor were important stakeholders and should share in the productivity gains

Items 1 through 7 were widely adopted in the United States, but 8 and 9 were not, especially in the early 1900s. The exception was Henry Ford who not only more than doubled the daily wage of his employees, but also added profit sharing to their compensation package. Typically the labor force was generally denied any gains in the benefits of the productivity increases. In fact, wages were lowered as labor content contributed less and less to the productivity gains. Taylor's work was attributed as the cause. Both he and his work were labeled oppressive, undemocratic, repressive, and a "stealing the soul" of the American worker.

Although scientific management as a distinct theory or school of thought was out of favor by the 1930s, most of its elements are important parts of industrial engineering, planning, scheduling, and management today. Taylor's reductionist approach was centered on replacing inferior work methods. The process he pioneered resulted in building skill sets into equipment, resulting in a less-skilled workforce. Businesses continued to emphasize engineering their way out of labor and high-wage environments. The result was mass automation. His early work led industrial engineering to focus efforts on the transfer of knowledge from skilled workers into tools, and the productivity gains have been huge.

The adoption of Frederick Taylor's principles of scientific management completely changed the landscape—it was a major shift. Any major shift in technology or markets renders the "old" relevant information and strategy "irrelevant." When such a shift occurs, previously successful actions and tactics no longer deliver the expected results. When a system constraint shifts, relevant information shifts and the existing policies, tactics, and execution metrics they are based on must shift in order to continue to drive results. The validity of the ROI formula as a measure of the total system, however, does not change. The importance of Brown's ROI formula, variable costing, and cost-volume-profit analysis can best be appreciated by understanding its role in the survival and success of General Motors. When a founding father of flow (Ford) faced off against a founding father of management accounting (Brown) for dominance of the automobile industry, relevant information won the day and modern management accounting was born.

Ford versus General Motors—A Lesson of Relevancy

By 1916, Henry Ford, the master of synchronized flow and F. Donaldson Brown, the master of relevant information, were on a collision course. In the 17 years from 1910 to 1927, their two companies set the course for business and supply chain management for the next 60 years. Although it is clear Ford did not invent the assembly line, he did perfect it. Ford also shared Taylor's view that work could be broken down into small standard tasks that would eliminate the need for craftsmen. Although Ford maintained he was not influenced by Taylor's work, the very industries he took his ideas from had all adopted Taylor's methods. All of the above had to have influenced the thinking of Ford and his drive to perfect the modern assembly line.

Prior to Ford, the automobile industry was a craftsman industry. Built by craftsmen and guided by German engineering, these cars were untouchable in terms of quality and unaffordable to any but the very wealthy. A set of craftsmen literally handcrafted a car in a room. Table 8.2 is a summary of the evolution of the American automotive industry, describing the history of modern production and management accounting.

Critical to Ford's success was the wave of vertical integration and the application of its methods to all the key components and parts production. An investment of $14.5 million between 1912 and 1914 to in-source created speed and flow to the assembly line. Ford was compelled to introduce a very high level of vertical integration because his needs ran far ahead of the ability of its suppliers to supply him in the volume and the rigorous quality he demanded. Centralization of manufacturing was a necessary step. Once this had been successfully achieved by 1913, the key problem became assembly rather than the manufacture of the millions of parts.

The high level of integration also made possible, remarkably low levels of inventory. This was because Ford *did not* focus on his machine shop and tool room efficiencies with a unit-cost perspective—there were no economic order quantities or artificial batch sizes to maximize individual resource efficiency based on unitized fix costs. He was focused on the system flow and velocity of contribution dollars. Ford focused on the flow of the right part, in the right quantity, to keep the assembly line flowing efficiently—system efficiency.

The moving assembly line was the endpoint of a policy of continuous improvements in plant and process machinery that had been underway since

Table 8.2 The Birth of the U.S. Auto Industry and Early Success of Ford

---

1899

The U.S. Census of Manufacturing recorded statistics on the automobile industry for the first time. Some 30 manufacturers had produced an estimated 2,500 automobiles and the industry stood 150th in the value of its products in the U.S. economy.

1900

By the end of 1900, there were 13,824 automobiles on the road in the United States and worldwide fewer than 10,000 cars were produced that year.

1901

Ransom Olds opened the first automobile factory in Detroit and produced 425 cars.

1902

At least 50 new firms began manufacturing automobiles in the United States.

1903

Henry Ford formed the Ford Motor Company in Detroit, Michigan, employing a dozen workers in a 250-square-foot plant to assemble automobiles.

1904

- The United States surpassed France, becoming the world's largest producer of automobiles.
- William Durant bought the bankrupt Buick Motor Car Company in Flint, Michigan.

1905

- Some prices: The new Brush motorcar: $800; a Ford Model K: $2800; a good horse: $150–$300
- United States produced 25,000 automobiles

1907

United States produced 43,000 automobiles

1908

- Buick Motor Car was the world's largest car company with 9000 cars sold.
- Durant formed General Motors, which was based on the combination of Buick, Cadillac, and Oldsmobile.
- The combined automobile companies had a net profit of $29 million.
- Between 1904 and 1908, at least 240 firms were established to manufacture automobiles in the United States.
- The first Ford Model hit the road. It sold for $825 and Ford sold 10,000 of them.

(Continued)

Table 8.2 The Birth of the U.S. Auto Industry and Early Success of Ford
(*Continued*)

1909

- Durant focused GM on growth by acquisition and sold 25,000 cars and trucks. Durant counted on expanding sales to provide capital for his acquisitions and growth.

-He did not count on Ford. Ford sold 20,000 automobiles and stopped production of higher priced models to concentrate on the Model T.

1910

- 458,500 registered automobiles in the United Stated. The industry is 21st on the list in value of product.

- Ford sold 320,523 Model Ts; opened his Highland Park plant; created a tool department focused on developing improved types of specialized machinery to support higher volumes of production and speed flow. The new specialized machines for producing parts were positioned strategically according to what they produced in order to produce parts as close to the assembly line as possible.

- Willy's – Overland produced 18,200, second behind Ford.

- GM's loss of sales to Ford forces a cash shortage and Durant is forced out as president. A consortium of banks raises $13 million in cash.

- Forced out of GM, Durant joined Louis Chevrolet to form the Chevrolet Motor Company backed by DuPont money. With the backing of DuPont and the success of Chevrolet, Durant began buying GM stock.

1911

More than 1 in every 5 automobiles on the road were Fords (22%).

1912

- Ford begins huge wave of vertical integration bringing all key manufacturing processes in-house; achieves huge productivity gains by applying its methods to all the key components.

- Between 1912 and 1914, Ford invested $14.5 million, of which only $350,000 was on the assembly line in the chassis shop. Continuous conveyor belts brought parts and materials to assembly areas.

- Highland Park production rate is 26,000 each month—without chassis assembly lines—300,000 Model Ts annually.

- Four years after its introduction, the Model T accounted for ¾ of all cars on America's roads.

Table 8.2   The Birth of the U.S. Auto Industry and Early Success of Ford
(*Continued*)

1913

- Ford was using moving lines to assemble subassemblies, magnetos, motors, and transmissions. Subassembly times dropped by 75% and subassembly units quickly outpaced the chassis assemblers.

- Ropes and windlasses were installed to pull the chassis through final assembly. By the end of the year the time to assemble a chassis was cut from 12.5 hours to 2.7 hours.

- Ford's labor difficulties become serious. Ford is experiencing a 300% turnover rate with his assembly line workforce and there is an attempt to unionize.

1914

January 5: Ford Motor Company announces a wage jump from $2.40 for a 9-hour day to $5 for an 8-hour day, with profit sharing.

- Ford installed a continuous chain to pull chassis through its plant. The moving line for final assembly was the last automation because it made no sense until all the bottlenecks in production and delivery of components had been cleared.

- Ford's continuous process improvement efforts resulted in about 15,000 separate pieces of machinery needed to support a moving assembly line for the Model T.

- The price of a Model T falls below $500 and annual sales rise to 248,000; more cars than all other American automakers combined.

Source: The history of American Technology—The Automobile Industry:
http://web.bryant.edu/~ehu/h364/materials/cars/cars%20_10.htm

the introduction of the Model T in 1908. Ford focused on the flow to the subassembly lines and then the flow to the chassis assembly lines and only then did he focus on automation for the chassis assembly lines. Ford understood that the rate of output was governed by the slowest task time in the assembly line process. The total efficiency of the system was governed by the rate of the slowest task time and it dictated the speed of the moving assembly line. A labor slowdown anywhere on the end assembly line would create a slowdown of his entire supply chain (a task below the capable standard) and undo the productivity gain of not just the assembly line but all of the capital spent on vertical integration. Doubling the wage of the assemblers was dwarfed by the gain in product speed, flow, and revenue opportunity.

Thus, Ford was not being altruistic. He understood that keeping his assembly line fully staffed with knowledgeable workers leveraged all of his invested capital. Based on the enormous productivity gains resulting from the introduction of moving assembly lines, Ford was able to launch the $5, 8-hour day. The move more than doubled the average wage, and Ford clearly understood that in doing so he was also helping to build the market for the Model T—a car everyone could afford, including his own workers. More importantly, he protected Ford from the labor unrest of the time.

Ford understood that his assembly line had to be fully manned as well as fully fed. He was not focused on least cost; he was focused on flow and revenue opportunity. He achieved the least cost because of his focus on maximizing system flow with the minimum invested working capital. Ford was wholly focused on flow as his number-one law of supply chain and it worked. "The increased velocity of throughput permitted Ford to reduce the price of his product until it was half that of his competitors, to pay the highest wages in the country for nonskilled work, and still to acquire a personal fortune that was larger than that of John D. Rockefeller or Andrew Carnegie."[7]

Ford drove this model to complete dominance. The year 1924 saw the second straight year of nearly 2 million vehicles. Over half of the cars in the world were Model Ts. Three years later, Ford closed the Model T factory, 100,000 people lost their jobs, and General Motors assumed a market lead that lasted 80 years. What happened?

How Did GM Beat Ford?

GM could not beat Ford on price or engineering excellence. Ford had already established total price dominance with both economies of scale and his vertical integration strategy. This strategy drove his inventory turns through the roof. Ford had no wait time—parts inventory did not sit. It was produced and delivered to the line immediately as needed. Ford used a central planning and logistical model to drive his vertical integration and ensure supply to the assembly lines. His activity and variable cost measurement models, standardization of components, processes, and products and

---

[7]Chandler Jr., Alfred D., *The Visible Hand: The Managerial Revolution in American Business,* Harvard University Press, 1977, page 280.

use of electric motors to reconfigure workflow allowed him to synchronize flow. The placement of parts production next to the lines ensured everything he needed for his cars from the raw materials on up was available at the rate the assembly line could pull. He truly mastered flow.

In 1914 Ford understood his dependency on increasing volume when he offered rebates to every Model T buyer if sales topped 300,000 in 1915. He passed his lower unit of cost based on volume increases to his customers after they were earned.

Besides the capital investment in vertical integration, the organization required coordination of huge numbers of activities and employees. Ford created a massive bureaucracy and the reward was massive economies of scale that allowed Ford to continually drive down the price of the Model T. His strategy was to create new market tiers and first-time buyers who could afford his product as the price declined. The downside was it required an unlimited pool of first-time buyers to continue to work. By 1919 it was clear that head-on price competition with Ford was futile, but it was also clear there was a hole in Ford's thinking. He never saw an endpoint to his strategy. He failed to have a strategy to change when the constraint shifted and consumer purchasing power and limited capacity no longer determined market share. He lacked the ability to measure and sense the market shift.

Ford was the undisputed master of flow because he understood how linear-dependent event systems behaved:

1. The slowest task governs the rate of any chain of dependent events.
2. The synchronization of activity to, through, and from those tasks creates system speed and velocity.

He understood what George Plossl clearly verbalized 70 years later as the first law of manufacturing (see Chap. 1): All benefits (ROI) will be directly related to the speed of flow of materials and information.

Ford was competing solely on price. Flow enabled a low price. As long as the market appreciated price above all else, Ford would win. This is a commodity-market strategy. Prior to 1919 and the formation of GMAC to provide financing for GM cars, the market behaved as a commodity market.

GM beat Ford by enabling a shift in the market through innovation in financing and marketing. Don't want to spend all of your cash? No problem! Want variety and choices? No problem! Their success was built on a system of relevant information for decision-making, pioneered by an engineer turned accountant, F. Donaldson Brown.

By 1919, GM was in the same trouble as every other car manufacturer competing against Ford. They were unable to maintain even their modest economies of scale and they were drowning in inventory. The introduction of GM's financing program in 1919, made possible by the DuPont backing, allowed GM to offer cars first to their dealers, who previously had only been able to purchase a few cars at a time. Secondly, they could tap into a new tier of consumers without meeting Ford's price of $350. The new financing program, although a move in the right direction, still required a $25 million cash infusion by DuPont to keep GM afloat. DuPont's 1920 management restructure brought Brown to GM and put Alfred P. Sloan in charge of finding a strategy to tackle Ford. Brown's first step was to get control of inventories and stop overproduction. The factories' output production schedule was tied to a strict monthly budget of materials purchased and labor and equipment scheduled to the rate of the forecast demand. He accomplished this with the variable budgeting technique he had developed at DuPont. Donaldson Brown tied flow to demand.

Although not evident at the time, the financing program along with Ford's failure to counter it, already set the inevitable in motion. The very consumers that were Ford's target market for the next round of price reductions were already becoming first-time buyers of GM cars. By the time the Model T fell below $300 in 1924, and to $200 in 1926, GM had already penetrated that first-time buyer market tier and was well on their way to perfecting the second tier of their market strategy, brand differentiation by style and by market price segmentation. The style strategy also addressed the second market shift Ford failed to identify, the used-car market. *GMAC and the financing innovation was again the brainchild of Donaldson.*

Even though Ford had continued to improve the quality, engineering, and features available on the Model T, he spent little time on the car's look. Changes to the Model T were based on functionality, not style. Black was the fastest drying paint, and color only added cost and time, both of which went the opposite direction of Ford's goal of offering the best *value* car. In 1923, continued spending and engineering delays on an innovative air-cooled engine for the Chevrolet were seriously eroding the brand's profit. Sloan ordered development to be sidetracked in favor of focusing efforts on giving their mass-produced Chevrolet the look of an expensive craft-built luxury car. The ensuing brisk sales of the new-looking Chevrolet convinced Sloan it was not necessary to lead in engineering, but merely to offer consumers better looking cars with more variety.[8]

The Birth of Push and Promote

Based on the success of the Chevrolet in 1923, GM's five brands were re-engineered to share parts. This cheapened production by increasing the economies of scale on which mass production relied. *The key point here is GM understood they should segment their market not their resources.* The different models and makes shared mechanical parts, such as transmissions and brakes, as well as their structural foundations, known as body shells. All of the GM models were built on three shells of different sizes. The same shell was made to look different by simply adding superficial features like fenders, headlights, taillights, and chrome trim that were unique to each model. Ford's basic black cars were eclipsed by GM's bright colored models.

By 1924 all of the divisions of GM were tied together with annual forecasts prepared for each division. These *division indicies,* as they were called, included not only purchases and delivery schedules for materials and capital equipment required and labor to be hired, but also estimated rates of return on investment and prices to be charged for each product. Prices, unit costs, and rates of return were all closely related to the volume permitted by demand. In drawing up these divisional indices, the staff computed the size of the national income, the state of the business cycle and normal seasonal cycles.[9] By 1926 GM's strategy had literally driven the Model T from the marketplace.

By 1927, the used-car market had grown large enough to become a competitive threat for first-time buyers. GM purchased and destroyed 650,000 used cars between 1927 and 1930 and prompted Sloan to introduce the annual model change. The model change of course relied

---

[8]Tough Guys and Pretty Boys: The Cultural Antagonisms of Engineering and Aesthetics in Automotive History, by David Gartman, http://www.autolife.umd.umich.edu/Design /Gartman/D_Casestudy/D_Casestudy3.htm
[9]Chandler Jr., Alfred D., *The Visible Hand: The Managerial Revolution in American Business,* Harvard University Press, 1977, page 460.

on aesthetics, not engineering advances. Every year the body style of all GM cars was changed slightly to give consumers the look of newness and progress and differentiate the owners who could afford new from those who had to settle for secondhand. However, underneath the dazzling new surfaces, the body shells and mechanical parts remained unchanged, often for decades. GM's strategies of focusing on financing and automotive style were responsible for propelling GM sales past Ford in 1927, when the Model T was discontinued due to plummeting sales.

GM used its five independent brands to segment the market based on price, features, and style. This was all made possible because of Brown's understanding of cost-volume-profit analysis and his new variable budgeting technique that allowed top management to understand the cost structure at different volumes. This allowed them to offer a price that was both competitive and profitable and the ability to judge investments against the return different investments created. Cost-volume-profit (CVP) analysis became an integral part of GM's management tool set. The idea of a target ROI compared against actual performance and the subsequent variance analysis to take corrective action was integral to GM's success. GM was able to establish price points necessary to sustain each brand's target profit and ROI after covering their fixed costs. *These management accounting innovations positioned GM to emerge as the dominant car maker and hold that dominance over Ford for 80 years.*

It is critical to note that Brown's management accounting tool set and all of his ROI equations and ratios as well as cost and variance analysis was based on variable costs and CVP analysis. *Unitized fixed costs never entered into their decision-making. GAAP accounting did not yet exist.*

Ford was not "wrong," but it can be argued he did not have a system that would point out when his productivity gains outstripped available demand and his continued investment focus and strategy ceased to generate a positive ROI. In fairness to Ford, surplus capacity had simply never been possible before. Like most truly outstanding innovators, Ford was stuck in his previous successful experience of how the world worked. He did not have a methodology in place to continuously evolve his organization. Despite numerous indications that his strategy was no longer effective, Ford continued to rely on economies of scale and productivity gains believing price alone would continue to be a competitive advantage in the marketplace. *Ford learned a hard lesson: flow only matters if the market wants what is flowing.*

By 1927 consumers were buying used GM cars and paying up to twice the price of a new Model T for GM's new "stylish" cars. Ford was chasing diminishing returns in productivity, the Model T market was saturated, and yet his management structure continued to focus on continuous improvement in productivity without paying attention to the very real change in market conditions his tremendous success had helped create. In fact, it can be argued Ford defeated himself because of the very nature of the structure he created and staying on a strategic path of diminishing marginal value. When Ford set his course in 1904, his goal was to make an automobile the common person could afford. Ford's goal was accomplished by 1919 and there was solid evidence that Ford's strategy was no longer effective. Not because he did not continue to lower the price, but because what people could afford and wanted had changed. The constraint governing automobile sales was no longer the cost of an automobile, nor were productivity gains the driver to success. Ford understood the importance of flow but failed to see the changes taking place in the market. His disconnect to the market disconnected flow from good ROI performance. Table 8.3 outlines the decisions and strategies of Ford and GM at the time.

Table 8.3 Ford's Battle with General Motors

1915

- Ford announced a rebate to all purchasers of Model Ts if sales exceed 300,000 and at year-end the Ford Company sent out checks totaling more than $15 million to 308,313 purchasers of Model Ts.

- Durant and DuPont and the Chevrolet company achieve a majority interest in GM and take over.

1916

- Alfred P. Sloan sold Hyatt Roller Bearing Company to GM and joined the GM organization as president of United Motors, a parts-manufacturing subsidiary.

- U.S annual automobile production surpassed 1 million units and over half were Ford Model Ts; the runabout sold for $345, and the touring model for $360.

1919

- GM mimicked Ford's production methods but needed a market to keep pace with its automation and productivity gains. General Motors Acceptance Corporation was formed and began financing both dealer inventory and consumer purchases of GM automobiles.

(Continued)

Table 8.3 Ford's Battle with General Motors (*Continued*)

1919

- Easy monthly payments allowed more people to buy GM cars than Ford.

- Ford, believing it was morally wrong, would refuse to offer a credit program to finance his cars until 1930. Instead, he focused on making his cars more affordable for first-time buyers.

1920

- Annual U.S. production of automobiles stood at nearly 2.3 million; the world production was just under 2.4 million.

- DuPont's directors invest an additional $25 million in a still struggling GM and Pierre DuPont becomes president.

1921

- Durant is forced out of GM and Sloan is promoted to vice president and tasked with devising a strategy to crack Ford's lock on the market.

- F. Donaldson Brown joins Pierre DuPont and Sloan at GM and becomes vice president of finance. He brings his decentralized management accounting practices, ROI formula, variable budgeting, and CVP analysis to GM's five car divisions.

- Sloan/Brown quickly rule out competing head on with Ford on price and settle on a strategy of style and variety.

- Ford's share of the U.S. market is 61% and GM's stands at 12%.

1922

- Under Alfred P. Sloan's leadership and with help from Brown's operational controls and financial models, General Motors sold 457,000 units and reported a profit of $61 million.

- GM management entrusted the Chevrolet line to a former Ford employee: William Knudsen. With assembly-line production, nine-year-old technology, a new body style, and Du Pont's colorful paints, Knudsen set out to make the struggling Chevrolet competitive with the Ford Model T.

- Durant Motors introduced its "Star" at $348 to compete with the Ford Model T. Henry Ford simply undercut Durant's price and drove him from the market proving head on price competition was futile.

1923

- Brown developed the financing mechanism that allowed DuPont to retain control of the GM investment.

- GM sold 800,000 vehicles and earned a profit of $80 million.

- Sloan orders engineering efforts on an innovative air cooled engine delayed in favor of a focus on giving mass produced cars the sleek look of craft-built luxury cars.

Table 8.3 Ford's Battle with General Motors (*Continued*)

1923

- Brisk sales of the new-looking Chevrolet convince Sloan that a strategy of offering consumers a better looking car with more variety trumps the need to compete with Ford in engineering.

- Well over half of the 3.7 million cars purchased in the United States were bought on credit.

- There were 108 U.S. firms manufacturing automobiles, of which 10 accounted for 90% of annual production. Ford accounted for just over 51% of cars produced.

- Thirteen million cars are on American roads.

1924

- Ford's production of Model Ts approaches 2 million for the second year in a row with a price point of $290. Over half of all cars in the world were Ford Model Ts.

- Brown was appointed to GM's executive committee and, working closely with Sloan, he refined his cost accounting techniques he had developed at DuPont.

- GM sold more than 800,000 vehicles and the Chevrolet, its lowest price model, sold for $525. It was advertised as the lowest priced quality automobile. A comparably "fully loaded" Ford was priced at $360.

1925

- The price of a Ford Model T Roadster dropped to $260 and Ford has 10,000 U.S. dealers selling Model Ts.

- Chevy sales soar to 341,281—all that Chevy could make.

1926

- In the United States, 43 companies were manufacturing automobiles. No new manufacturers succeeded in entering the industry after this date.

- U.S. production of automobiles reached 4 million.

- GM introduced the Pontiac brand, which was actually just a rebranding of Oakland. The 1926 Pontiac models offered significant luxury and performance for the low price of $825 and they were a runaway sales success. Within 12 months, a total of 76,742 units would be produced. The car ranked as America's 13th bestselling automobile for the year.

- GM's Art & Color section was formed under the guidance of Harley J. Earl. It was the first department at any automaker devoted solely to "styling" cars. Its work would give each GM division a design identity. It answered two competitive threats: Ford and the used-car market.

- Unheard of just a decade earlier, credit sales of automobiles had become the industry standard. Installment purchases accounted for more than two-thirds of all new-car sales. With the automobile leading the way, credit purchases of expensive consumer goods (e.g., home appliances) were becoming a way of life for Americans.

(Continued)

Table 8.3 Ford's Battle with General Motors (*Continued*)

1926

- Facing increasingly stiff competition from Chevrolet, a new Ford Model T with a self-starter sold for just $350. Ford was struggling to keep up. Ford insisted its restyled 1926 Model T was "totally new," but in fact, the utterly obsolete design was on its last legs.

1927

A bad year for car manufacturers. Effectively, the phenomenal market growth of the past 20 years had ended. The market had shifted to replacement of used vehicles. The trade-in and used-car market had come to play an increasingly important role in automobile sales.

- Last year of production for the Ford Model T. Sales dropped to one-third the 1926 level. Even at prices around $200 Ford couldn't maintain sales, and by this point consumers were buying 3- and 4-year-old used cars rather than a new Model T because they perceived the used cars as better value—with more features. By the time Ford shut down production of the Model T, the company had sold 15,458,781. The only car to outsell the Model T would be the Volkswagen Beetle.

- There were more than 20 million cars on the road in the United States; automobile sales made the industry the leading American industrial sector in value of product. Automobiles ranked third on the list for value of product exported. There was one registered automobile for every 4.5 Americans, and 55% of American families owned a car.

- The American market of first-time buyers of new automobiles had reached saturation. For the first time, sales of replacement vehicles surpassed the combined total of sales to first-time purchasers and purchasers of multiple vehicles (fleets).

- Second-hand automobiles in excellent condition sold for about the same price as a Model T and the "stylish" Chevrolet cost just $200 more.

- Closing down production of the Model T meant 100,000 workers lost their jobs. Ford closed production and began the design and tooling changeovers that produced the new Model A.

- Between 1927 and 1930, Chevrolet purchased and destroyed 650,000 used cars in an effort to prop up the market for new vehicles.

- GM introduced the annual model change to differentiate itself and compete against the used-car market.

- For the first time, GM sales surpass those of Ford and would remain there for 80 years.

1928

GMAC signs its four-millionth retail contract.

Systemizing the Management of ROI

First, the transformation from craft production to mass production was completed. This created a market based on economies of scale and scope. It gave rise to giant organizations built upon functional specialization and minute divisions of labor. Economies of scale were produced by spreading fixed expenses, especially investments in plant and equipment and the organization of production lines, over larger volumes of output, thereby reducing unit costs. Economies of scope were produced by exploiting the division of labor. The development of formalized management information and accounting systems to measure performance, direct tactics, evaluate strategies and management performance provided the ability to organize and control these entities. Production time standards, material standards, variance analysis, operating ratios, flexible budgeting, and CVP analysis were developed and widely adopted across all businesses. Management accounting and industrial engineering came into being as distinct management disciplines and provide the framework today of all reporting, scheduling, and planning functions.

By the 1930s, Ford's standardized commodity product, and his direct planning and control system had been rendered obsolete by innovations in marketing and organization at General Motors. His logistical system was being emulated by his competitors, but the focus for investment and improvement was not productivity gains or even product improvements. These innovations were implemented by Alfred P. Sloan, who is best known for the multiproduct, or M-form, organizational structure, in which each major operating division serves a distinct product market.

When Sloan took over GM in the early 1920s, it was little more than a loose confederation of car and car-parts companies. Sloan repositioned the car companies to create a five-model product range from Chevrolet to Cadillac and established a radically decentralized administrative control structure. Ironically, within each of its operating divisions, GM was organized and operated like Ford—or any other mass-production manufacturer. In this system, assemblers were as interchangeable as parts. The mass-production system rested on the presumption that activities should be simplified to the nth degree and controlled from above, engineering and administrative functions delegated to staff specialists, and the exercise of judgment passed up the managerial ranks.

GM's operating divisions—the five automotive divisions, the divisions making components, and those making refrigerators, air conditioning, locomotives, and so on—were managed entirely by the numbers from a small corporate headquarters, using the DuPont system of financial controls, devised by Donaldson Brown who became GM's chief financial officer. Under this system, each division kept its own books and its manager was evaluated in terms of a return-on-assets target. If the numbers showed that performance was poor, it was time to change the division manager. Division managers with consistently good numbers got promoted, ultimately to headquarters. Short-run coordination between GM's five automotive divisions and the divisions making components (e.g., Fisher Body or Delco-Remy) was achieved via buyer–seller relationships. Longer-run coordination was achieved via the first modern capital budgeting system (flexible budgeting based on variable costing) used in the United States, and also devised by Brown.

The rules to support this approach are solidly in place in every major corporation worldwide, and the DuPont ROI model is still used today. In fact it is the basis for all modern financial ratio analysis used by both analysts and public accounting firms. Unfortunately the cost components of Brown's DuPont formula were destined to be appended in 1934 when the SEC was legislated into being and GAAP was born. As previously explained in Chap. 3, a regulating body known as the Securities and Exchange Commission (SEC) and the AICPA are responsible for the alteration of Brown's ROI formula. The SEC was brought into existence by legislation with the intention to protect the public and the financial market from fraudulent misrepresentation, by companies, regarding the reporting of annual and quarterly financial statements. GAAP was born and product costs were defined as both variable costs and fixed costs for external financial reporting. Although there was and is no intention to distort companies' internal management information, the distortion has occurred over time.

When companies adopted Taylor's product routings with time and material standards, the focus was on time and usage as inputs. Actual to standard variances were examined on an exception basis to understand process deviations as well as to focus improvements. The costing system was intended to capture variable costs directly associated with the product. Time standards and routings were also used to provide the information necessary to plan, purchase, and schedule resources. The shift to mass production and the lack of information technology created a middle management charged with recording transactions and keeping accounting records. As pressure increased to reduce the cost of middle management and given the limitations of early computing power, managerial accounting information took a backseat to financial accounting information, to the point of nearly disappearing from the radar. The necessity to follow GAAP absorption costing to support external reporting regulations and tax code overrode the priority of variable product costing in support of Brown's ROI formula, variable budgeting and CVP analysis. Thus, most companies today cannot differentiate between relevant information for decision-making based on the change in cash flows and product cost information used to make a management decision.

Companies would be better off today if the rules Ford and Brown operated under were the only rules used to define relevant information for decision-making. Their rules were based on variable cost and fixed cost behavior consistent with economic CVP models explained earlier and a flow-centric efficiency strategy. In Chap. 10, we will take these same principles and update them to complexity theory rules for nonlinear systems, rules that fit today's complex supply chain environments and rules that are the foundation for smart metrics and the cornerstone of the 21st century breakthrough in management accounting.

The Emphasis on GAAP and the Rise of Unfocused Improvement and Outsourcing

As previously discussed in Chap. 3, the nail in the coffin for providing variable costing information arrived in the early '80s with the widespread adoption of MRP II and its ability to automatically roll up all manufacturing costs into one top-level number. Unfortunately, that number is a fully absorbed unit cost and it is being used inappropriately at all levels of the organization to drive decisions that are at best suboptimal and at worst disastrous.

The deep truth of cost-centric efficiency is at the core of both companies' outsourcing decisions and local cost efficiency improvement actions and projects. Companies *believe* in a linear, Newtonian view of the world. Every efficiency gain, anywhere, translates to an increase in system productivity because Newtonian math is additive. Additive math works only for an independent or single-event system. An efficiency gain at any one resource translates to an increase in system *speed* but does not change the *rate* of output (governed by the slowest unit) or the truly variable cost of producing any product. Maximizing local efficiencies does not maximize *system* efficiency. Carnegie, DuPont, Ford, and Brown understood this. Unitizing a fixed cost never entered their paradigm. There were no incentives to artificially build inventory. They were solidly focused on flow-centric efficiency and ROI.

Unfortunately companies act as if they believe that GAAP standard absorption unit cost is a true representation of cash flow. The assigned standard fixed dollar cost rate, coupled with our Newtonian view of the world leads managers to believe that every resource minute saved anywhere is the equivalent of a dollar cost savings to the company. GAAP unit costs are used to estimate both cost improvement opportunities and cost savings for batching decisions, improvement initiatives, and capital acquisition justifications. In reality the *cost* being saved has no relationship to cash expended or generated and will not result in ROI gains of the magnitude reported. Cost savings are being grossly overstated. Today it is not uncommon to find plants running with burden rates over 1000 percent.[10] Unfocused machine or labor efficiencies are meaningless and even counterproductive. They are completely overstated through the use of GAAP absorption unit costing. A decrease in fixed costs will simply not happen due to a time savings at any single resource area. The total fixed cost spend will remain the same, regardless of what the unit fixed cost does.

Previous efforts of the 1920s through today, to reduce direct labor, have been incredibly successful. The ratio of labor dollars to overhead investment dollars, as a percent of product cost has totally flipped since the turn of the twentieth century. Touch time labor today averages less than 7% of the product cycle time (meaning parts and product are waiting to be worked on 93% of the time) and direct labor accounts for less than 10% of total product cost in most manufacturing organizations. *The opportunity for supply chains is not labor or machinery productivity gains but to focus on getting rid of the "wait time" that impedes flow.*

---

[10]Miller, Jeffrey G., and Thomas E. Vollman, "The Hidden Factory," *Harvard Business Review,* September 1985.

CHAPTER 9

A Case Study—The Boeing Dreamliner

Companies need to be looking at every decision from the standpoint of ROI opportunity for the whole system not just the cost component. Companies have chased the focus on labor reduction past the point of diminishing returns and into the realm of negative returns. One of the most damaging GAAP misuses has been the use of full absorption unit costing to drive outsourcing decisions. Bad outsourcing decisions were first driven by comparing unitized labor and overhead costs to the outsourced part purchase price. If neither labor nor the overhead cost assigned to the product cost will actually be reduced when a part is outsourced, then there is no real dollar savings.

Developing economies, with "very" cheap labor have invested in both the latest technologies and infrastructure to take advantage of the offshoring opportunities created by the knowledge transfer from skilled labor to technology. This has resulted in global oversupply and excess capacity across the world, as well as very complex global supply chains.

This transfer of knowledge to technology has made offshoring feasible. However, companies have flawed assumptions about relevant cost information making offshoring both attractive and "profitable," which have led to a series of very bad decisions to go offshore. There are some very good reasons to go offshore. Those reasons have everything to do with strategic contribution and nothing to do with unit cost savings. When a company doesn't understand the relevant information to make a good offshoring decision has been illustrated in a very well-publicized case that continues to make headlines as this book is being written.

The biggest irony with outsourcing and offshoring is that the wait time in the supply chain increases with outsourcing and offshoring. The increased lead time requires greater investment in inventory to buffer between the customer's tolerance time (little) and the time to order, make, and ship the product to the point of its pull (a lot more). None of this increased "investment" is considered as relevant information in the decision to offshore. More often than not, the fixed costs remain unchanged and the remaining products are left to absorb all of the overhead (a lower volume of work over the same fixed costs). As a result, more products are outsourced because it is *cheaper* to buy them than to make them. Ultimately the organization's abilities to design and manufacture are simply gone. In many ways this is the story of lots of manufacturing leaving the United States and Europe.

Businesses are waking up to the negative effects of losing control of their supply chain. The trend to outsource operations has leveled off and actually decreasing in the United States compared with five years ago. According to a recent study published by DVV Media Group GmBH, 2013 (www.bvl.de); Trends and Strategies in Logistics and Supply Chains Management—Embracing Global Complexity to Drive Market Advantage states: "Outsourcing operational activities has previously been emphasized but is decreasing in the United States. Today less that 30% of logistics activities is outsourced compared with 39% in Germany and 50% in China. One reason for the absence of increased outsourcing may be due to dissatisfaction with the outsource provider performance." Examples abound, but at the time we are writing this book, the Boeing Dreamliner is grabbing big headlines. It illustrates all of the lessons of a complex supply chain being managed with cost-centric efficiency rules and the rules of complex systems not being understood by top management.

The Boeing Dreamliner

On July 8, 2007, Boeing rolled the Dreamliner out of a hangar in Everett, WA before a crowd of thousands. The aircraft's design attributes included its highly fuel-efficient composite exterior, powerful engines, and game-changing interior. Boeing had already taken 677 preorders, making it the first commercial plane to pass the 500 mark for orders ahead of first delivery. At the time, the company said it would put the first Dreamliner in the air by September of that year, 2007. The first passengers would be transported in the new plane by May 2008. From there, things unraveled quite quickly.

As we are finishing this book (March 2013) the 50 Dreamliner jets Boeing has delivered are grounded due to the battery packs spontaneously combusting and there appears to be no quick fix on the horizon. On January 16, 2013, the Dreamliner was grounded by the FAA for all U.S. airlines. Japanese, Indian, and European aviation authorities followed quickly with their own grounding orders. Boeing continues to assemble Dreamliners but they cannot deliver them. They cannot fly them to their customers even if their customers wanted them. They are running out of places to park them. Airlines are lining up to demand that Boeing pay them delivery penalties, lost compensation for canceled flight schedules, and loss of revenue from their existing fleet of Dreamliners. The current scenario is a lengthy recertification with the FAA to recertify flight worthiness.

To further complicate the situation, in July 2012, Boeing top management replaced Jim Albaugh, the Boeing Commercial Airplanes (BCA) CEO who had regained both the trust and respect of his engineers. He is also credited with rescuing the Dreamliner. He was replaced with Ray Conner, who was more aligned with top management thinking. This strongly signals a return of BCA leadership, to the strategy and tactics of the pre-Albaugh days that led to the current situation. At a time when they critically need the help of their engineers, they face a strike initiative vote with members of the Society of Professional Engineering Employees in Aerospace (SPEEA), a possibility that was unthinkable 8 months ago under Albaugh's leadership.

Regardless of the outcome in the next six months, we have tremendous faith that the very talented people at Boeing will resurrect the Dreamliner. The question isn't: Will the Dreamliner be salvaged? The question is: What will Boeing executive management and the rest of industry learn from the experience?

As demonstrated by the Dreamliner, Boeing appears stuck in the conventional cost-centric deep truth, which has been described at length in the previous chapters. For the Dreamliner, Boeing executives chose a dramatic departure from Boeing's previous successful airplane launches because Boeing was no longer "Boeing." This shift in strategy can be traced to the 1997 merger with McDonnell Douglas that saw Harry Stonecipher and Phil Condit ascend to lead the new Boeing. Ironically, Stonecipher has a degree in physics and Condit was by all accounts a talented engineer. The Dreamliner was to be their legacy, and they are largely responsible for the shift to a cost-centric efficiency strategy that has been at the heart of the culture clash since the merger.

The following is an excerpt from James Surowiecki's article, "Requiem for a Dreamliner?" (*The New Yorker*, February 4, 2013):

> The Dreamliner was supposed to become famous for its revolutionary design. Instead, it's become an object lesson in how not to build an airplane.
>
> To understand why, you need to go back to 1997, when Boeing merged with McDonnell Douglas. Technically, Boeing bought McDonnell Douglas. Richard Aboulafia, a noted industry analyst with the Teal Group described it as follows: "This is when McDonnell Douglas in effect acquired Boeing with Boeing's money. McDonnell Douglas executives became key players in the new company, and the McDonnell Douglas culture, averse to risk and obsessed with cost-cutting, weakened Boeing's historical commitment to making big investments in new products. After the merger, there was a real battle over the future of the company, between the engineers and the finance and sales guys." The Dreamliner's advocates came up with a strategy that was supposed to be cheaper and quicker than the traditional approach: outsourcing. And Boeing didn't just outsource manufacturing parts; it turned over the design, the engineering, and the manufacture of entire sections of the plane to some 50 "strategic partners." Boeing itself ended up building less than 40 percent of the plane.
>
> This strategy was trumpeted as a reinvention of manufacturing. But while the finance guys loved it—since it meant that Boeing had to put up less money—it was a huge headache for the engineers.

Boeing and McDonnell Douglas appeared to be a match made in heaven. Boeing had a strong commercial air program but was weaker on the space and defense side. McDonnell Douglas had a strong space and defense program but was quickly becoming irrelevant on the commercial side. McDonnell Douglas leadership speculated that they would have to invest $30 billion in development just to get back in the commercial air game.[1]

In the years following the merger, with Condit driving strategy and Stonecipher focused on operations, the relationship between Boeing

---

[1]Callahan, Patricia, "So Why Does Harry Stonecipher Think He Can Turn Around Boeing?" 2004, Chicago Tribune Company, LLC.

management and its unions had deteriorated dramatically and there had been no spending on new airplane programs. It culminated in a contentious strike with SPEEA in 2000 with outsourcing as one of the major issues. SPEEA represents roughly 23,000 Boeing employees and had traditionally viewed itself as both loyal and important stakeholders in the Boeing business. Stonecipher's outspoken determination to break the family culture of Boeing and make it a business had not been well received.

In addition, Stonecipher's strong support of cost-cutting initiatives was widely believed to be flawed and in fact credited with taking the company in the wrong direction both in quality and overall system efficiency. With Stonecipher's exit, there was hope that the company would move forward with the 7E7 (the original designation until being changed to 787 in 2005) and the oversight and management of the program would remain inside Boeing.

Stonecipher retired in June 2002, and analysts, union leaders, as well as many workers believed that the company would get back on track as the leader in commercial aviation. Stonecipher had been widely criticized as so focused on profits, shareholders, and the bottom line that he had lost sight of what made Boeing great: being a company not afraid to take huge financial gambles, such as with the development of the 707, 747, and 777.

> While Airbus got bigger, "Boeing stagnated." They failed to formulate a strategy that could keep up with an emboldened Airbus, as Boeing fell behind in both technology and manufacturing efficiency during the '90s. Boeing, once the manufacturing marvel of the world, now spent 10 to 20 percent more than Airbus to build a plane. The loss in market share—from nearly 70 percent in 1996 to roughly 50 percent in 2003—marked an astonishing reversal.[2]

After Stonecipher's retirement there was a concerted effort by Boeing senior management to take back their company. In an article published March 9, 2003, in the *Seattle Times,* "Boeing buzzes about 'source' of work," the concern over just how much of the work on the 7E7 would be outsourced exposed a focus much larger than the loss of jobs in Seattle. While Stonecipher had retired, his legacy was very much alive and engineering management continued to struggle to communicate their concerns. Surprisingly, it was the work of Dr. John Hart-Smith, a Boeing

---

[2]Holmes, Stanley, "Boeing: What Really Happened," *Business Week,* December14, 2003.

senior technical fellow that ended up making headlines in Seattle. His 2001 paper on outsourcing, presented at a Boeing conference, became so widely read in Boeing it ended up coming to the attention of the *Seattle Times*. The struggle for control of managing the 7E7 moved to a public arena, when the article was published by the newspaper:

> A controversial internal paper warns that excessive outsourcing could lead to the loss of the company's profits, its core intellectual assets, and even its long-term viability. The most important issue of all, is whether or not a company can continue to operate if it relies primarily on outsourcing the majority of the work that it once did in-house. The author is John Hart-Smith, a senior technical fellow at the Phantom Works research unit in Southern California and one of Boeing's most eminent engineers.
>
> Boeing officials declined to respond directly to Hart-Smith's paper but agreed to discuss their outsourcing strategy: Boeing, they argue, bring together the best technology from partners around the world. And as it faces relentless competition from European rival Airbus, Boeing has no choice but to manufacture planes more efficiently—which means letting suppliers build more complex components.
>
> Hart-Smith, however, argues that even if outsourcing makes sense in some circumstances, Boeing's reliance on it is excessive. "Outsourcing all of the value-added work is tantamount to outsourcing all of the profits," his paper concludes. "It is time for Boeing to reverse this policy."
>
> His argument has created a lot of buzz inside the company, according to a senior program manager who asked not to be named. But it has been largely dismissed by Wall Street. Chris Mecray, an analyst with Deutsche Bank, called Hart-Smith's paper "more of a rant than anything."
>
> "Hart-Smith's McDonnell Douglas heritage is crucial to his paper, which repeatedly cites the case of the Douglas DC-10 as the apotheosis of outsourced aircraft manufacture. The excessive use of subcontractors on the DC-10, Hart-Smith asserts, eroded most of the commercial-airplane unit's profits and thus its ability to fund new jets. By the time it merged with Boeing, McDonnell Douglas had all but exited the commercial-airliner business in favor of building military aircraft.

Hart-Smith has not been cleared to discuss the details of his paper. Boeing allowed him to provide only a written response to emailed questions about his background.

"My motive," he wrote in a note accompanying his responses, "was simply to save Boeing suffering the same fate as befell Douglas. I became concerned about seeing too many of the same policies being advocated."[3]

Hart-Smith's paper was anything but a rant. His background was extremely relevant in making the argument against excessive outsourcing in the Dreamliner program. Before the 1997 merger, Hart-Smith had spent 29 years with McDonnell Douglas. Hart-Smith's comment points out how both the financial institutions and the top management of companies are laboring under the same flawed assumptions connected to the deep truth of the cost-centric efficiency strategy and financial statement information.

Hart-Smith's paper is not only logically sound; it is based on very relevant experience. It is a clear statement of the predictable negative effects when irrelevant costs, information, and performance metrics are used. It is not surprising that an engineer of his background and experience would come to the same conclusions as both Henry Ford and F. Donaldson Brown. We have read Hart-Smith's paper and there are six key points to be summarized in relation to this book.

1. Selective outsourcing can be beneficial to all concerned as a supplement to sales activities outside a company's fixed cost volume range. It is an additional cost to receive the additional sales and contribution margin that could not be satisfied otherwise. This is known as offloading a constraint.
2. As a tool for reducing costs, outsourcing is most commonly flawed because of the misleading cost accounting practice of including irrelevant information from unitized fixed overhead in comparing part cost. These "costs" will not be reduced; the potential "savings" is distorted and overstated.
3. In the event the fixed overhead in both skill sets and machinery are eliminated to reduce the net asset investment of the company as a strategy to improve RONA (return on net assets, another version of ROI) the company will place itself at the risk of not being able to compete in

---

[3]Gates, Dominic, "Boeing Buzzes About 'Source' of work," *Seattle Times Aerospace reporter*, March 9, 2003.

the future when new products requiring these scarce skill sets and machinery are needed. As described previously, this was the exact place the commercial air division of McDonnell Douglas found itself in during the mid-1990s, requiring Stonecipher to seek a merger with the Boeing company. This assessment on the need for periodic investment spending on new products is validated by Richard Aboulafia's analysis and commentary on why the 2003 decision to invest in the 7E7 was critical for Boeing to remain competitive in the commercial airplane market.

4. Ignoring the increase in overall costs from both additional tasks and additional investment in inventory to protect, synchronize, and speed the assembly, flow must be assessed both from a cost standpoint as well as a risk standpoint. Carnegie and Ford both understood this point, which is the reason they pursued strategies of vertical integration and insourcing. They did not want to lose control of their supply chain and risk losing control of both their costs and ultimately their market. The additional tasks are not trivial and in the case of a new airplane program they are immense.

   a. If outsourcing is to be employed it is critical that detail parts and subassemblies be designed with that purpose in mind to avoid the situation whereby major subassemblies do not fit together at final assembly and then require rework and/or redesign, increasing both the costs and cycle time by orders of magnitude.

   b. To address this requires considerable *additional* upfront effort in planning and the prime contractor must provide on-site quality, supplier management, and technical support.

   c. Off-site production increases the total span of time and transportation costs. In order to compensate for variation in time, the prime contractor must invest in additional inventory of all of the subassemblies and parts to protect the assembly line from disruption. Inventory must be sized to protect the assembly pull against the reliable time to replenish from the source. The longer the replenishment cycle time, the larger the inventory on hand must be. Hart-Smith said it well, "Inventory costs accrue throughout the span time between installation (no matter by whom) and sale of the (final) product, *not* throughout the span time for final assembly alone! 'Creating' that same reduction by merely transferring work to suppliers is *not* a system efficiency gain!" In other words the total cost of the investment of the supply chain in time expansion will end up being passed through to the end consumer. The end cost of the product will go up.

5. The risk-sharing partners by definition do not recoup their investment in the product program until their parts are used in the final assembly. Suppliers that are on time and on specification are punished when other parts of the supply chain do not perform. Hart-Smith made this point in his paper: "The prime manufacturer can never exceed the capabilities of the least proficient of their suppliers." And in a *risk-sharing partnership* neither can the rest of the supply chain partners. They will not recoup their investment any faster than the product can flow to the market. This is a physical law and the basic tenet of the Theory of Constraints as well as Plossl's first law of manufacturing.

6. "RONA is an excellent goal by any standards. Only when it is converted into a performance metric do its devastating effects become apparent." What Hart-Smith is discussing here is exactly our discussion in early chapters regarding the interdependence of ROI on five key objectives: sales, costs (spending), quality, due date performance, and investment (inventory and capital). They do not move in isolation of each other and any attempt to push RONA to a local level for tactical decisions results in a compromised tactical decision and action with a deleterious effect on the whole system of which RONA and ROI are the measurement after the fact.

In his 2001 paper, Hart-Smith went on to predict several outcomes if Boeing applied their strategic partner outsourcing policies to the new 7E7 program.

1. They would be unable to make the pieces fit together in final assembly.
2. They would lose control of their supply chain and be forced to send their engineers and technical experts into the field to solve the problems their suppliers would encounter.
3. They would be forced to salvage the program by buying and or bailing out their strategic partners by paying them for the research and capital costs.
4. Both the cycle time to produce and total cost for the plane would increase by orders of magnitude.
5. The product would require redesign and rework that would be unknown until after all certification tests were done and known.

In December 2003, Stonecipher was called out of retirement and was again at the helm of Boeing after Condit's forced resignation. Despite the marketing hype and strong suggestions that Boeing would launch both the 7E7 (the Dreamliner) and the 747-8, Condit had not pulled the trigger. The board was expected to decide whether to commit to the development of the 7E7 jetliner by the end of the year. The 7E7 would be Boeing's first new airplane in a decade. The decision would determine Boeing's commitment to the commercial airplane market. "This is really a pivotal moment," says Aboulafia. "Failure to invest in the 7E7 could mean the beginning of the end for Boeing's storied airplane business."[4]

In January 2004, the board approved the 7E7, soon-to-be-named the *Dreamliner*. Simultaneously, Boeing made permanent the appointment of James Bell as chief financial officer. Bell had been acting CFO since the firing of his predecessor Mike Sears a few months earlier. Unlike Sears, who had an engineering background, Bell had always been a numbers guy, with a 31-year track record in corporate finance. Together with Mullaley, the then-CEO of BCA, they put together a plan to spread the risk of development costs for the 7E7 among "50 strategic outsource partners" as a way to make the 7E7 more palatable to their board and the financial community. This strategy was sold as a reinvention of manufacturing instead of the same "old" McDonnell Douglas strategy and approach.

Stonecipher was wholly focused on cost reduction as *the* strategy. Between 1998 and 2001, when he oversaw Boeing operations, his unrelenting drive to cut costs—and his sharply critical, outspoken approach—made him widely unpopular with rank-and-file employees and with unions. Stonecipher went on to defend Boeing's continued globalized outsourcing. In a *Seattle Times* interview on August 2, 2004, Stonecipher was very clear in discussing Boeing's prospects, global competition, local jobs, and the "strategy."

> We're going to continue to reduce costs. We have a plan, and [Commercial Airplanes CEO Alan Mulally and his team] are developing it very nicely. It's going right down the road, and where we've done it, it works fine. This is not about monetizing assets. It's not about getting rid of people. It's about lowering the cost of the product.
>
> Every time we've done this, we get significant cost reductions in the hardware. That's what we want.[5]

Boeing executives mandated the outsource tactics for the Dreamliner for the simple reason they believed it would maximize their return on net

---

[4]Holmes, Stanley, "Boeing: What Really Happened," *Business Week,* December 14, 2003.
[5]Gates, Dominic, and David Bowermaster, "Blunt Boeing CEO Bullish on Company's Prospects," *Seattle Times,* August 2, 2004.

assets (RONA) by spreading the capital investment and the risk necessary for the development of the 787 across a broad range of suppliers—the 50 "strategic partners." They believed that the 787 could be reduced to 50 different smaller projects, discrete independent events, and more easily managed, hence development would be faster. By structuring the suppliers' arrangements such that their development costs could only be recovered over the life of the program, Boeing originally believed its total investment in R&D would be under $6 billion.

They appeared to ignore many of their own engineering experts and external supply chain experts who tried to explain the flaws of the approach. They wholly discounted the risks associated with their decision to not just outsource manufacturing but the engineering as well. They received congratulations and accolades from the financial community for their "new strategy" to spread the risk of the invested capital. The expectation of success would be measured in the long run by RONA but the market rewarded them immediately. Table 9.1 shows a timeline of the progress (or lack of progress) of the Dreamliner. Remember the objective was flight certification by 2008 and Boeing believed that it would invest less than $6 billion.

In February 2010, the Boeing company released its financial statements for 2009 showing a write-off of $2.7 billion out of the capitalized R&D in the inventory/program account for the 787 as a year-end adjustment to its 2009 financial statements. Per the annual report, they elected the write-off because three of the first six Dreamliners were not salvageable due to reengineering.

In January 2011, Boeing announces its seventh Dreamliner delay and schedule slide. This set off another barrage of articles focused on the Dreamliner and for the first time a public admission by BCA CEO Jim Albaugh that they are re-examining and adjusting some outsourcing decisions. Not surprisingly, it brings Dr. Hart-Smith and his original white paper back under scrutiny and numerous articles cite his report from 2001.As of February 2011, he was batting a thousand with a better track record as a prophet than Nostradamus. *Seattle Times* reporters conducted interviews with both Dr. Hart-Smith and Jim Albaugh for their article, "A 'prescient' warning to Boeing on 787 trouble" (*Seattle Times* business staff, *Seattle Times*, February 5, 2012). Hart-Smith had retired from Boeing in 2008.When he was asked, "Where did a structures engineer get that kind of expertise?" Hart-Smith replied, "It's common sense." Table 9.2 continues the saga of the Dreamliner's progress.

Table 9.1   The Dreamliner Timeline, January 2003–January 2011

January 2003	Boeing sets up a team of executives to design the new plane.
June 15, 2003	The company says the plane will be called the *Dreamliner* after 500,000 people from more than 160 countries voted on the name.
December 16, 2003	Boeing announces that its board of directors has given final approval for the plane. The company starts taking orders from airlines.
July 26, 2004	Japan's All Nippon Airways becomes the launch customer for the 787 with an order for 50 planes.
December 2004	Boeing ends 2004 with 56 orders for the plane. Its goal had been 200.
December 2005	Boeing ends 2005 with a total of 288 orders for the plane.
June 2006	Assembly begins at the company's plant in Everett, Washington.
June 2007	Engineers assembling the first plane find a 0.3-inch gap between the nose-and-cockpit section and the fuselage section behind it.
July 8, 2007	The world gets its first glimpse of the Dreamliner as it is paraded in front of a crowd of 15,000. (The plane turns out to be just a hollow shell, rushed together for the event.)
September 2007	Boeing announces the first of many production delays, starting with a shortage of bolts and problems with flight-control software.
2008	Boeing announces four more delays during the year. Among the reasons: a 57-day machinists strike, problems with improperly installed fasteners, and trouble with the company's global supply chain.
	SPEEA adopts media tactics to highlight the "disaster of the 787 program" keeping the spotlight on the outsourcing problems plaguing the company. (Source: June 10, 2011, Leadership Conference, Seattle Airport Hilton, Ray Goforth, Comments to the SPEEA council. Full text of the speech is available on **www.speea.org.**)
June 2009	Boeing reports 59 cancellations for the 787.
July 2009	Boeing spends nearly $1 billion to acquire two plants from Vought Aircraft Industries; the supplier had been responsible for large parts of the 787 air frame and continual part shortages and delays.
June 23, 2009	Boeing announces another delay, citing the supply chain.

Table 9.1 The Dreamliner Timeline, January 2003–January 2011 (*Continued*)

December 15, 2009	The first 787 test flight leaves from Paine Field, adjacent to Boeing's factory in Everett, Washington.
February 2010	The Boeing company correctly elects to write-off $2.7 billion out of the capitalized R&D in the inventory/program account for the 787 as a year-end adjustment to its 2009 financial statements. Per the annual report, they elected the write-off because three of the first six Dreamliners were not salvageable due to reengineering.
November 9, 2010	During a test flight, a 787 loses electrical power after fire breaks out in an electrical control panel. Test flights are delayed six weeks.
January 2011	In an appearance at Seattle University, Boeing Commercial Airplanes CEO, Jim Albaugh, had some comments about the 787's global outsourcing strategy as well as the lessons learned from the disastrous three years of delays on the 787 Dreamliner. One bracing lesson that Albaugh was unusually candid about: the 787's global outsourcing strategy—specifically intended to slash Boeing's costs—backfired completely. "We spent a lot more money in trying to recover than we ever would have spent if we'd tried to keep the key technologies closer to home," Albaugh said. Boeing was forced to compensate, support, or buy out the partners it brought in to share the cost of the new jet's development, and now bears the brunt of additional costs due to the delays. Albaugh said that part of what had led Boeing astray was the chasing of a financial measure called RONA, for return on net assets. "We went too much with outsourcing," Albaugh said in the interview. "Now we need to bring it back to a more prudent level." Albaugh balked at going that far, referencing Hart-Smith's assertion that Boeing should keep most of the work it has traditionally done in-house. "I haven't said keep *most* of the work in-house," Albaugh said. "I still believe we need to make sure we try to access the best technologies and capabilities that are available around the world."(Source: "A 'prescient' warning to Boeing on 787 trouble," Seattle Times business staff, Seattle Times, February 5, 2012.)

Source: Unless other sources are cited, a timeline of Boeing's 787 Dreamliner, Associated Press, January 25, 2013.

Table 9.2 The Dreamliner Timeline, February 2011–February 2013

February 2011	Boeing is more than $2 billion over budget (120%) and 3 years late. Journalists, aerospace analysts, and Boeing's customers are all raising the same three questions posed and answered by Peter Cohan:
	1. Specifically, did Boeing outsource too much of the Dreamliner's components to other companies in other countries?
	2. Will the 787's outsourcing problems persist?
	3. What might this mean for airlines, passengers, and investors in Boeing stock?
	The short answers are: yes, probably, and it's too early to tell.
	(Source: Cohan, Peter, "Boeing's Dreamliner Delays Outsourcing Goes Too Far, "*Daily Finance,* January 21, 2011; http://www.dailyfinance.com/2011/01/21/boeing-dreamliner-delays-outsourcing-goes-too-far/)
June 2011	Ray Goforth, executive director of SPEEA, announces a dramatic change from the negotiation tactics of 2008 and proposes adopting binding binary arbitration. He summarizes the progress Boeing management has made and the shift in the relationship as follows: "That was then. This is now. 2012 is not 2008. In the interim since 2008, many things have changed at the company. Boeing has embraced a return to engineering excellence. The company has admitted its outsourcing mistakes and is actively reversing many of them. The company has built upon the original SPEEA-sponsored crosstalks to start fostering a culture that surfaces problems early. BCA president Albaugh has brought a different management style . . . one that recognizes that labor unions can be used by management as an early warning system to identify problems.
	(Source: June 10, 2011, Leadership Conference, Seattle Airport Hilton Ray Goforth-1.jpg, Comments to the SPEEA council. The full text of the speech is available on www.speea.org.)
August 26, 2011	The Federal Aviation Administration and the European Aviation Safety Agency certify the plane to carry passengers.
September 25, 2011	Japan's All Nippon Airways takes delivery of the first 787.
October 26, 2011	Three and a half years behind schedule, the first paying passengers step aboard the plane. The four-hour charter flight on ANA goes from Tokyo to Hong Kong.

Table 9.2   The Dreamliner Timeline, February 2011–February 2013 (*Continued*)

July 2012	July 28. Debris falls from a 787 engine during a test, sparking a grass fire at South Carolina's Charleston International Airport.
	Boeing's top management replaces the engineer CEO Albaugh, who was credited with turning around BCA and making it better again. They replaced him with a non-engineer CEO, Ray Conner.
January 7, 2013	The battery pack on a Japan Airlines 787 catches fire after the flight landed at Boston's Logan International Airport. Passengers had already left the plane, but it took firefighters 40 minutes to put out the blaze.
January 11, 2013	The FAA launches a review of the entire plane, even as top transportation regulators insist it is safe.
January 16, 2013	An ANA 787 makes an emergency landing after pilots are alerted to battery problems and detect a burning smell. ANA and Japan Airlines ground their entire Dreamliner fleets.
January 16, 2013	The FAA grounds all 787s flown by U.S. airlines. Japanese, Indian, and European aviation authorities follow with their own grounding orders.
February 2013	February 8. SPEEA and Boeing not in agreement. In a last-ditch attempt to avert a strike by his engineers and technical workers, Boeing CEO Ray Conner (Albaugh's replacement) sent a letter to all SPEEA members asking them to accept the company's offer. With strike authorization ballots arriving in the mail this week, Conner is hoping to avoid a strike when he needs all of his SPEEA members on the job. He wrote it's time to come together as a team to work for the company's future.
	"Nobody wins in a strike," he wrote. "While hurting Boeing and our employees, it would also impact our customers who've put their trust in Boeing's people and products. It's important that we protect our competitiveness in the longrun, even if that means some short-term pain."
	SPEEA executive director Ray Goforth said the letter was not well-received by his members. "Boeing corporate is trying to take advantage of that by telling people the company's in trouble, just accept this terrible offer and help us get these 787's back in the air," he said.
	Goforth said it's strange that Conner would be getting involved at this late date. "He's played no role in these negotiations . . . "
	(Source: Sullivan, Chris, "Boeing takes SPEEA union appeal directly to workers," MyNorthwest.com, February 8, 2013.)

Not surprisingly the January 16, 2013, grounding of the Dreamliner brought the program management and the supply chain outsourcing of the 787 back to the forefront as journalists and analysts scrambled to make sense of both the past and the future of the Dreamliner.

Repeatedly over the past 18 years, one name continues to appear in articles and commentary from both industry analysts and business journalist alike: Richard Aboulafia. His commentary and insight are sought after with good reason. His newsletters are pointed but balanced with both compelling strategic assessment as well as numbers. He is a respected and longtime aerospace industry analyst and is frequently cited as an aviation industry authority by trade and news publications as well as television news and radio programs.

The following is an excellent summary on the Boeing Dreamliner from Aboulafia's January 2013 newsletter, which can be found online at: http://www.richardaboulafia.com/shownote.asp?id=373:

> I don't like change. But every so often, I'm forced to re-examine my long-held beliefs. This is one of those moments.
>
> During my 25 years as an aerospace industry analyst, Boeing looked more competitive than Airbus. My job is to forecast and provide recommendations to industrial and financial clients, and Boeing almost always delivered better results. When random Euro-sycophants accused me of being "anti-Airbus," (as if it were a cultural/national bias) I merely offered to discuss numbers (i.e., financial returns and benefits to industrial partners). They'd get that confused deer-in-headlights look (which confirmed that many people in this business don't quite understand market economics), and sadly mope away.
>
> I expected this routine to continue for the next 25 years. Yet times have changed. A client recently showed me some jarring numbers. As of January 25, EADS's stock has returned 137% or 7.1% annualized since it was listed in July 2000. In the same period, Boeing has returned 117%, or 6.4% annualized. This trend, of course, is quite recent. Boeing stock has been impacted by the 787's problems. It has significantly underperformed broader U.S. stock averages with a total return of about 1% (including dividends). EADS, on the other hand, has been soaring

and is up 36% since the start of 2012, with about 20% of that in the past month. Before that steep rise, Boeing's stock would have still performed better than EADS's. Underperformance or outperformance obviously depends on the starting and end dates.

Yet EADS has risen while battling its own serious headwinds over the past year, and while generally underperforming Boeing in revenue and profit. Boeing jetliner output rose 47% by value in 2012 over 2011, while Airbus rose just 12%. Boeing's profit margins are considerably better than EADS's. With the 777 and 787, Boeing has a better twin-aisle product line, while Airbus is stuck with that A380 albatross. EADS/Airbus still has serious governance and ownership issues. EADS has seen more savage home market defense budget cuts than Boeing has. It also has seriously underperformed Boeing in defense export markets, as evidenced by the F-15's continued long run and Eurofighter's string of campaign failures. The battle over share prices and investor returns shouldn't be a battle at all. Boeing should be way ahead.

How to explain this disconnect? I've got a theory. The jetliner business is a duopoly, and the jetliner market is growing faster than any other manufacturing industry. Investors want to be part of it. And Boeing is starting to scare them. After all, Boeing's stock price hasn't been clobbered by the 787 problems; rather, it has simply stayed stagnant. Airbus/EADS, by contrast, has been benefiting from a strong jetliner market without any program disasters. In other words, it isn't that there's something particularly right with Airbus's strategy. Rather, there's something wrong with Boeing's strategy, and their execution.

Let's review. Last summer, Boeing's top management axed the engineer CEO who had been turning around BCA and making it better again. They replaced him with a non-engineer CEO. Then, management got into a confrontation with the engineer's union (which may also partly be the union's fault, but it's not a battle management can afford right now). Then Chicago put off the very promising 777X until the next decade, which, from a customer perspective, might as well be an indefinite postponement. These moves were on top of a 787 development model that de-emphasized in-house engineering and relied on industry partners for much of the development work.

Since the 787 appeared to be out of the woods, and the 777X was put off until the next decade, Chicago likely didn't think it needed much from engineers. Then that damn 787 battery thing happened. Oops. Back in Seattle, engineers, represented by a disgruntled union and forced to report to multiple layers of non-engineer management, are working overtime on the problem, but after several weeks, nobody appears to be close to a solution. As this is written, the likely outcome is a six- to nine-month grounding (due to the need for recertification).

This terrifying state of affairs for the Dreamliner, of course, was merely background for Boeing's fourth quarter earnings call this month. The 787 fiasco wasn't discussed, except that (a) the investigation was continuing and couldn't be discussed and (b) 787 production was continuing full speed ahead, despite uncertainties about what needed to be done for the battery system, or any other aspects of the plane's design. If these planes being built need major retrofit work in the future, well, that's for the engineers to worry about.

Meanwhile, there was no contrition or soul-searching on the call about how the 787 could have gone this wrong, or what could be done within the company to make it right (once again, 787 program analysis was left to the journalists). Instead, the call emphasized some impressive sales and profit numbers. It was like a farmer showing off a great crop, but not mentioning that the tractor just broke, he fired the mechanic, and outsourced tractor maintenance to Bolivia. And that customers for next year's crop had been promised penalty payments if the farm didn't deliver.

Chicago's view of engineering, as seen in management changes, union negotiations, product launch decisions, and design outsourcing moves, is that it's a secondary consideration, far behind financial and market considerations such as return on net assets (RONA). But clearly this strategy of downplaying engineering is starting to have a deleterious effect on the company's financial performance, at least in terms of equities returns relative to the competition. Sure, investors may be scared by the high compensation costs associated with the 787's woes.

But it's also possible that investors may be getting spooked by a company that seems to lack a proactive approach for dealing with a serious crisis. Even when the 787 gets back to service, it may face further difficulties. There's also the likelihood that Boeing may be returning to the bad old days of 1998–2003, when it spent next to nothing on new product development.

In other words, Boeng's problem isn't just that the engineers have been nudged aside by the bean counters. It's that the bean counters need to rethink the way they manage the company. Until that changes, investors may continue shifting their focus towards Airbus's virtues, particularly if Airbus continues to emphasize spending on new products.

Nothing makes a more compelling statement for the necessity of change in current supply chain management strategy and tactics than Boeing's Dreamliner. Sadly, it was predictable and it was predicted repeatedly by both their internal and external experts as early as 2000. The roots of the problem tie directly into Aboulafia's assessment regarding the "bean counter mentality," and where that mentality originated. The 1996 merger with McDonnell Douglas and Harry Stonecipher's influence with Phil Condit who was focused on an acquisition strategy to diversify Boeing put the efficiency and least cost strategy firmly in the driver's seat. It overtook the company culture and sanity. Boeing is no stranger to large variances to plan but the point of variances analysis is to point out you have a problem and need a course correction. Large aggressive projects have variances. The outsourcing trend Boeing followed was simply one of the inevitable outcomes of using the cost-centric efficiency strategy to manage a complex supply chain, a strategy Stonecipher followed at McDonnell Douglas to its detriment as well. This strategy first raised its head under Condit in 1995 but under Stonecipher it appeared to become a full-blown mode of operation. The Dreamliner was a perfect storm and Boeing's current dilemma has three key lessons that every company should pay attention to:

Lesson 1: Creating Unnecessary Supply Chain Complexity

Many top executives in large organization have little understanding of the realities of the supply chain they are managing and the technical realities of the projects and products they are bringing to the market.

The increase in complexity in the technology of the Dreamliner coupled with the complexity Boeing added to its supply chain by increasing not only outsourcing content (30 percent of content – 5 percent for the 747) but also created a level of supply chain complexity impossible for them to control let alone synchronize and manage. They had no visibility and no proactive management of their supply chain. They were forced into a reactive mode beginning in 2007 with their attempt to assemble the first Dreamliner. It was the second, and third, and fourth before they got the major components to fit. Costs spiraled out of control as they did everything necessary to rescue the program including flying hundreds of their internal experts to supplier sites, purchasing underperforming suppliers and repeated redesign.

Figure 9.1 is a recreated depiction of the change in supply chain structure for the 787 from previous airplane programs as depicted and presented by Christopher Tang and Joshua Zimmerman of the UCLA School of Business.

To reduce the 787's development time from six to four years and development cost from $10 to $6 billion, Boeing decided to

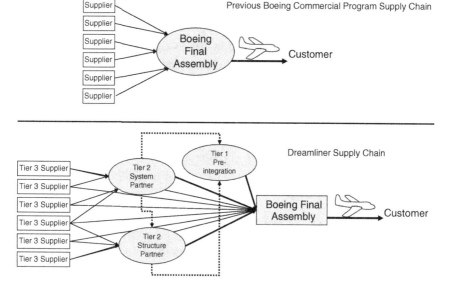

Figure 9.1 The Change in Boeing's Supply Chain Configuration

develop and produce the Dreamliner by using an unconventional supply chain new to the aircraft manufacturing industry. The 787's supply chain was envisioned to keep manufacturing and assembly costs low, while spreading the financial risks of development to Boeing's suppliers. Unlike the 737's supply chain, which requires Boeing to play the traditional role of a key manufacturer who assembles different parts and subsystems produced by thousands of suppliers [top of Fig. 9.1], the 787's supply chain is based on a tiered structure that would allow Boeing to foster partnerships with approximately 50 tier-1 strategic partners. These strategic partners serve as "integrators" who assemble different parts and subsystems produced by tier-2 suppliers [bottom of Fig. 9.1]. The 787 supply chain depicted in Fig. 9.1 resembles Toyota's supply chain, which has enabled Toyota to develop new cars with shorter development cycle times.[6]

Their conclusion was Boeing dramatically increased the complexity of their supply chain for the 787 without assessing or understanding the increased risks of supply, process, management, labor, and ultimately the risk to their market. Furthermore, they concluded Boeing did not have the expertise, visibility, or supply chain management tools and structure to undertake such a dramatic shift. The lack of any risk assessment and mitigation strategy and no proactive supply chain management visibility forced Boeing into a continuing series of reactive responses resulting in cascading delays backward and forward through the supply chain. The study was conducted in 2009 and Boeing was still two years from a successful first delivery. The following excerpt is the conclusion of Tang and Zimmerman's article.

Besides the need to perform due diligence in key supplier selection to ensure that the selected supplier has the requisite capability and the commitment for success, a company should consider cultivating stronger commitment in exchange for accurate information in a timely manner. Overly relying on IT communication is

---

[6]Tang, Christopher S., and Joshua D. Zimmerman, "Managing New Product Development and Supply Chain Risks: The Boeing 787 Case," *Supply Chain Forum An International Journal*, Vol. 10, No. 2, 2009, page 77

highly risky when managing a new project. To mitigate the risks caused by partners further upstream or downstream, companies should strive to gain complete visibility of the entire supply chain. Having clear supply chain visibility would enhance the capability for a company to take corrective action more quickly, which is more likely to reduce the negative impact of a disruption along the supply chain. See Sodhi and Tang (2009b) for a discussion of the importance of timely response to mitigate the negative effects of supply chain disruptions . . .

Boeing took something inherently complex and made it an order of magnitude more complex. Furthermore, while making the supply chain more complex they appeared to use no decoupling and dampening and feedback mechanisms to break the predictable increased supply chain dependent event variation. With the increase in dependencies and connections without decoupling and dampening and feedback mechanisms they literally created a perfect recipe and environment for the bullwhip effect. The speed of flow of relevant information and materials was a trickle.

Boeing's management thinking, business rules, and information system tools were completely inadequate for the task. They were and are based on old science, as are 99 percent of businesses today, but few have their level of complexity. Their Newtonian approach to attempt to control and manage both risk and complexity was to break the supply chain apart into independent events and manage and control each piece as if they didn't have to fit together at the end. To say it didn't and doesn't work is an understatement at this point. Over the last 20 years, companies' complexities have increased and their thinking and tools have been increasingly failing. They have sought to push their problems and risk onto their suppliers. The trend has been away from the vertical integration.

Vertical supply chain integration not only makes controlling the system easier it also allows for dramatic profit potential in industries with after-market service parts. Markets and profit multiples are won and lost on the companies' abilities to manage and service their spares business. This is a key point Dr. Hart-Smith made in his paper—a significant portion of the profit margin in an airplane program is in those spare parts. The average piece of equipment with a life longer than 10 years will pay its original selling price again over its 10-year life. Airplanes have a practical life of 20 years.

Mandated maintenance and the necessity to purchase parts from only Federal Aviation Agency approved suppliers mean that spare parts in the commercial aircraft industry are a gold mine.

Lesson 2: Industries with Program Accounting Should Beware

The ability to derive relevant information at all levels becomes more challenging when given the use of Program Accounting, an aerospace GAAP staple. Program accounting is only used for firms with very large R&D investment and capital requirements to startup a product line. Overhead and indirect labor spend as a ratio to direct labor dollars are very large and hence the misplaced focus on cost-cutting and reduction efforts aimed at indirect labor—outsource engineering—save money and drive up RONA. Program accounting is simply a version of GAAP absorption costing on steroids. The shift in supply chain design driven by a cost-centric efficiency strategy obscured and distorted the ability to judge the progress and status of the program's RONA.

The problem with program accounting is you get to pretend your RONA is great and following your plan right up until after you start to ship the first plane. It is only then a company understands the magnitude of the capitalized overhead and R&D each plane will have to absorb over the program life. The 787 has become a very expensive airplane and "it ain't over yet." Program accounting for financial statement purposes is a valid approach for financial statement presentation for a company like Boeing, and is based on a core GAAP principle known as matching. It allows firms that must make very large capital and or R&D expenditures, to produce the first shippable product and then spread the investment over all products shipped over the life of the program. It requires companies to estimate the life of the product and the number of products that will be shipped over that timeframe.

Boeing has a track record of actual volume sales far exceeding both the number of planes as well as the number of years the program remains active (e.g. the Boeing 737). This is a very good thing as it means they have consistently outperformed their projected payback and return on their airplane programs. It also causes a very odd anomaly only seen in program accounting, negative inventory. When the number of planes exceeds the original program life, program accounting continues to allocate the standard full absorption overhead cost to each and all additional aircraft to be expensed as the cost of goods sold.

So even though the total investment has been recovered and the inventory balance of the program startup dollars capitalized has been fully allocated out—zero dollars—credits are still applied against the program inventory balance sheet account and debited to the cost of goods sold account for every additional airplane sold. This results in a negative dollar value program inventory account balance. This of course dramatically understates net income and cash flow for mature programs and overstates net income and cash flow for startup programs such as the 787 Dreamliner. But it does fulfill the GAAP matching principle designed to spread the expected development costs over the life of the airplane program and match the cost of each plane to the sales revenue in the period the sale takes place.

An outside analyst or investor cannot possibly know the cost associated with any single airplane program, as all program inventories are collapsed into a single number on the balance sheet. The negative inventory balance from all programs that have exceeded their program life are netted against new programs and existing programs with life remaining. It is a black hole; what has really been spent on the Dreamliner is openly speculated to be somewhere between $12 billion and $18 billion. Dramatically different than the $6 billion originally estimated when Boeing management sold their innovative supply chain risk sharing methodology to Wall Street.

The last thing to note is the decision to expense the cost of the first three Dreamliners rather than capitalize them was not really a choice. We are certain Boeing's auditors (Deloitte) required the write-off as the variances were both large (almost half of the original expected development cost) and unusual. The first three Dreamliners produced could not be sold because they couldn't get the pieces to fit together. It was determined they could not be salvaged for sale due to extensive redesign and rework after testing. This was exactly Dr. Hart-Smith's point and prediction in his 2001 paper. In Boeing's 2010 annual report and financial statements, Boeing referenced the 2009 write-off and stated that additional write-offs might be necessary in 2011 if additional technical or supply chain issues arose. The $2.7 billion is not considered part of the program overrun—it was expensed. Today (June 2013) Boeing still does not know the total cost of this airplane.

Boeing management appears to have refused to acknowledge the reality of what was happening and what was not working or they simply were so focused on the individual part and supplier problems they never saw the systemic problem until it was too late to prevent the write off of four entire airplanes and years of program delay. Either way it does not reflect well on upper management or the strategy they chose.

Lesson 3: Understand and Value Your Organization's Human Capital

Top management risks the loss of confidence of both their stakeholders and their stockholders if they do not understand the meaning or value of all of their assets, let alone how to maximize the return of their assets. Boeing executives appear to have discounted the value and the talent of their workforce and its place in the past and future success of Boeing.

Dr. Hart-Smith, a Boeing Senior Technical Fellow, used Brown and Ford's definitions of flow, synchronization, and relevant cost and benefit information coupled with constraint-based principles in his internal paper on Boeing outsourcing. The Hart-Smith paper is worth reading in its entirety and is available on the web. The journalistic shortcuts to explain it are reduced to out-of-context sound bites, but the exposure it received and continues to receive has endured for 13 years for good reason. He was right and it is indisputable. Seldom is there the opportunity to follow and compare such well-documented and specific predictions from an inside expert, over the span of time necessary to test the conclusions…and both it and him were available to Boeing management from the beginning. The utilization of this knowledge could have dramatically changed the 787 program for the better.

When a company doesn't understand the value of their people or recognize them as an asset, they don't know how to drive return from their total organization. In almost every organization the authors have experienced, the collective knowledge and intuition in a company's work force is one of its most valuable assets. Treating human capital as a commodity is a poor business decision especially "knowledge intensive" human capital.

In the New Normal human capital has become as important to business as it was before the Industrial Revolution. Chapter 10 will explain the importance of human capital to supply chain management given they are complex adaptive systems.

Authors' Note:

Living in the Pacific Northwest, the authors have a heightened level of interest in Boeing as well as direct access and contact with the company, even nondisclosure agreements in place. Given our backgrounds, our areas of expertise, and this access and connection we feel we are uniquely qualified to assess Dr. Hart-Smith's paper. Boeing invented program accounting. In particular, Debra Smith is very familiar with both Boeing and program accounting, as she spent the first years of her career in public accounting, on the audit staff of Touche Ross, now Deloitte Touche, assigned to the Boeing account and the Commercial Airplane Division. During her tenure as a professor at the University of Puget Sound and the University of Washington, she sat on advisory boards for Boeing and served with Boeing executives. When Dr. Noreen and Debra Smith received a grant from the IMA Foundation for Applied Research, the research affiliate of the Institute of Management, the IMA assigned Bob Miller as our grant handler and he worked with us throughout the project. This was 1993 and Bob Miller was the corporate controller for the Boeing company as well as the chairman of the IMA Foundation's Project Committee. In 1998 Smith was invited to provide a day-long workshop for the controllers of all the airplane programs (727, 737, 747, 757, 767, and 777) to help them assess both their costing and decision-making system. They got it but it was clear to them and they made it clear to Smith that Stonecipher was taking BCA down the opposite road from what she was proposing. Boeing engineers, their managers, and operation managers continually filled our workshops and conferences, and constraints-based thinking was not only popular but repeatedly proposed to upper management without gaining major traction.

In 2000 prior to the strike, we were contacted by Charles Bofferding, the then executive director of SPEEA, in hopes that we might be able to help them better communicate to top management their concerns regarding both the dangers of a strategy to increase outsourcing as well as how to debunk the cost savings numbers for both the proposed outsourcing and their current process improvement initiatives focused on cost cutting. All of these used financial absorption costing as the basis for the tactics and local performance metrics, which are outlined succinctly in Dr. Hart-Smith's paper. We had been recommended to him by senior managers in the Boeing company in both engineering and operations based on both our familiarity with Boeing, our expertise in management accounting, constraints management principles, and supply chain expertise. There was hope of averting the strike and of generating a rational dialogue of the risks and assessment of the outsourcing issues. As evidenced by the quality of the Hart-Smith's paper, it was not that Boeing did not have internal expertise in these areas, but they hoped outside "experts" would be helpful. Unfortunately for all, there were no opportunities for constructive dialogue. The train had left the station and we were not on it and neither were Boeing's team of experts. Boeing had entered the era termed by Boeing insiders as management by magical thinking—no logic and deep denial of the reality of the outcomes.

The deep truth of cost-centric efficiency is at the core of both companies' outsourcing decisions and local cost efficiency improvement actions and projects.

1. Companies and their managers believe in a linear, Newtonian view of the world. Every efficiency gain, anywhere, translates to an increase in system productivity because Newtonian math is additive. Additive math only works for an independent or single-event system. An efficiency gain at any one resource translates to an increase in total system speed but does not change the rate of output (governed by the slowest unit) or the truly variable cost of producing any product.

   Today parts and product spend 93 percent of their time waiting. A 10 percent speed increase in a resource that only accounts for some small percent of the total potential touch time is an infinitesimal gain in the system flow and speed.

2. Companies believe in the linear Newtonian approach to problem solving, the reductionist approach. The best way to gain control is to break everything into individual pieces and manage each optimally.

   The more complex the system the more quickly this approach will lose control. This approach actually makes the outcome of the project and/or product less predictable in terms of quality and reliability, elongates the market lead time, and escalates the total cost.

3. Companies act as if they believe in the GAAP standard absorption unit cost as a true representation of cash flow. The assigned standard fixed dollar cost rate, coupled with our Newtonian view of the world, leads managers to believe or act as if they believe that every resource minute saved anywhere is computed as a dollar cost savings to the company. GAAP unit costs are used to estimate both cost-improvement

opportunities and cost savings for batching decisions, improvement initiatives, and capital acquisition justifications.

In reality the "cost" being saved has no relationship to cash expended or generated and will not result in an ROI gain of the magnitude reported. Cost savings are being grossly overstated. Today it is not uncommon to find plants running with burden rates over 1000 percent. (Source: Miller, Jeffrey G., and Thomas E. Vollman, "The Hidden Factory," *Harvard Business Review*, September 1985.)

Maximizing local efficiencies does not maximize system efficiency. Carnegie, DuPont, Ford, and Brown understood this. Unitizing a fixed cost never entered their paradigm. They were solidly focused on flow-centric efficiency and ROI.

CHAPTER 10

Complexity Science and Supply Chains

In previous chapters, we have touched on the dramatic shift in the environment and circumstances that define modern supply chains. The understanding has evolved from linear structures to complex systems and most recently to a solid case for logistic systems as complex adaptive systems (CASs). This change is a result of the natural sciences shift in its view of the universe from solely the Newtonian or linear paradigm to accepting the addition of chaos theory, complexity theory, and complex adaptive systems theory or nonlinear paradigms, as an important part of the foundation for understanding natural order in the universe. These theories, although still emerging, have been accepted in mainstream science for the past several decades and have only recently become a hot and applicable topic to the social and applied sciences, such as economics and supply chain management.

Understanding the ramifications for business and supply chain management of the new rules of complex adaptive systems provides the framework for defining what and where to find relevant information. A bridge to the hard sciences and their more valid statistical models for complex supply chain behavior is the key. The new rules provide managers a solid framework to understand:

▲ How their system behaves
▲ What to measure and why
▲ When to take action and what result to expect
▲ Where to look for confirmation of the expected result

All of the above are necessary to provide relevant information to manage a supply chain based on flow through their system. Tactics, reporting, and measures aligned to a flow-centric strategy unleash the power of people to continuously adapt and improve their results.

The purposes of this chapter are to:

1. Confirm the validity and place of complexity and complex adaptive system science to explain and manage supply chains and the bullwhip effect.
2. Better explain the current Newtonian science rules and how they are used to derive the cost-centric efficiency strategy policies, tactics, measures, and information reporting common in today's supply chains.
3. Explain the new rules based on the laws and behavior of complex adaptive systems (CASs).
4. Provide a bridge from the CAS rules to derive the flow-centric efficiency strategy, policies, tactics, measures, and information reporting to manage today's supply chains.

Managers have little hope of understanding or improving the systems they manage if they do not step into the twenty-first century. It's analogous to bringing a knife to a gunfight. The following quote from Murray Gell-Mann, recipient of the 1969 Nobel Prize in physics and a distinguished fellow and co-founder of the Santa Fe Institute for the study of complexity science, puts context to the importance of understanding complex systems and the "laws" that govern them.

> Today the network of relationships linking the human race to itself and to the rest of the biosphere is so complex that all aspects affect all others to an extraordinary degree. Someone should be studying the whole system, however crudely that has to be done, because no gluing together of partial studies of a complex non-linear system can give a good idea of the behavior of the whole.[1]

The Newtonian Way

The world doesn't work the way we think it does or the way we have been taught it does. "Old science" still dominates the common views of reality.

---

[1]Gell-Mann, Murray, *The Quark and the Jaguar*, Henry Holt and Company, LLC, 1994, p. XI.

A scientific revolution began in the seventeenth century with Sir Isaac Newton's development of the calculus and laws of classical mechanics. Thereafter, scientists viewed nature from a profoundly different perspective. For the first time, Newtonian physics made it possible for scientists to determine the dynamics of bodies by simple equations. From the late-seventeenth century until the early twentieth, the Laws of Motion and other linear, mechanical principles discovered by Isaac Newton dominated the understandings of science and filtered down into every aspect of the Western world. This view of reality penetrated our education system, our culture, our language, our organizations, and our management practices so completely that it became taken for granted. The result has been that interdependence and interconnectedness were deemed less and less important and essentially ignored.

Most people are not even conscious that they are using what is called the mechanical view of reality and Newton's four golden principles, when they think, talk, and act. This view of reality assumes:

1. *Order:* given causes lead to known effects at all times and places. Things happen because something causes them to happen (cause and effect).
2. *Reductionism:* We can understand what happened by reducing things to their components or parts and examining those parts. There are no hidden surprises; the whole is the sum of the parts, no more and no less. Small changes lead to small effects and large changes lead to large effects.
3. *Predictability:* The universe is orderly, follows natural laws, and works like an incredibly complicated machine. Once global behavior is defined, the future course of events could be predicted by application of the appropriate inputs to the model. The inputs always equal the outputs. These models can be optimized.
4. *Determinism:* processes flow along orderly and predictable paths that have clear beginnings and rational ends. The belief that the past completely determines the future and everything would be and is predetermined—no chance, no choice, and no uncertainty.[2]

---

[2]Rihani, Samir, *Complex Systems Theory and Developmental Practice*, Zed Book Ltd., 2002, page 67.

Newton's four rules are responsible for three prevalent beliefs and they are the foundation for the cost-centric efficiency strategy used to manage supply chains today:

1. The best way to manage people is to break the organization into functions and organize them into a clear structure. Then control their actions with clear directions regarding their specific function. This follows Frederick Taylor's method of work practices and Ford's bureaucratic model.

2. The best results are obtained when work is streamlined at each unit to be as efficient as possible, with a minimum of wasted effort, producing the most output in the least amount of time. The lean-machine strategy will optimize any system output.

3. All cost structures are linear, additive, and divisible and can be directly associated with time increments, linearly, additively, and divisibly. A penny saved is a penny earned—everywhere and all of the time and GAAP unit costs are real pennies. The sum of all of the "cost savings" will fall to the bottom line. If every function in the organization performs to their best average the system will perform to its best average. The sum of all the average best times will equal the average system's best time.

None of the above is true in a nonlinear system. The beliefs all have unintended negative consequences for the complex organizations that follow them.

The Push Beyond Newton

Even from the very beginning, there were difficulties and anomalies that were left out under the universal order paradigm. Isaac Newton and Christiaan Huygens did not agree on the nature of light. Is it a wave or a particle? We now know it is both, but these difficulties remained politely below the surface.[3] It was unacceptable to challenge the core principles in academic circles and universities; in fact it was career-ending. They were often explained away as unimportant anomalies that would be resolved by the next wave of emerging fundamental laws. However, by the early twentieth century they could no longer be ignored.

---

[3]Ibid.

Henri Poincaré (1854–1912), the supreme physicist and mathematician of his age, was one of the first to voice disquiet about some contemporary scientific beliefs. He advanced ideas that predated chaos theory by some 70 years (Coveney and Highfield, 1996: 169). Later, Einstein's (1879–1955) theory of relativity, Niels Bohr's (1885–1962) contribution to quantum mechanics, Erwin Schrödinger's (1887–1961) quantum measurement problem, Werner Heisenberg's (1901–1976) uncertainty principle, and Paul A. M. Dirac's (1902–1984) work on quantum field theory all played a decisive role in pushing conventional wisdom beyond the Newtonian limits that had dominated for centuries.

These scientists, all Nobel laureates, set in motion a process that eventually transformed attitudes in many other science disciplines, but it was an uphill battle. The same inertia that is currently blocking adoption of new thinking and methods in the business community blocked the scientific community as well. Most people do not like challenging and rethinking their deeply held beliefs. It threatens how we define ourselves as successful, logical, and ultimately our worth.

Complexity

What if we could use what we already know and find a way to keep what is valid as the launching platform to an even higher level of understanding and performance? Intuitively most managers know there is something "wrong" or something "more," but there is no clear verbalization of what is wrong and no clear path to more. Understanding complex systems and its application can provide both. The new science of complexity evolved out of a field of study known as *chaos theory* that really gained ground in the scientific community with a discovery in 1961 by Edward Lorenz, a mathematician and meteorologist at the Massachusetts Institute of Technology.

> On a winter day 50 years ago, Edward Lorenz, SM '43, ScD '48, a mild-mannered meteorology professor at MIT, entered some numbers into a computer program simulating weather patterns and then left his office to get a cup of coffee while the machine ran. When he returned, he noticed a result that would change the course of science.
>
> The computer model was based on 12 variables, representing things like temperature and wind speed, whose values could be

depicted on graphs as lines rising and falling over time. On this day, Lorenz was repeating a simulation he'd run earlier—but he had rounded off one variable from 0.506127 to 0.506. To his surprise, that tiny alteration drastically transformed the whole pattern his program produced, over two months of simulated weather.

The unexpected result led Lorenz to a powerful insight about the way nature works: small changes can have large consequences. The idea came to be known as the "butterfly effect" after Lorenz suggested that the flap of a butterfly's wings might ultimately cause a tornado. And the butterfly effect, also known as "sensitive dependence on initial conditions," has a profound corollary: forecasting the future can be nearly impossible. Yet his insight turned into the founding principle of chaos theory, which expanded rapidly during the 1970s and 1980s into fields as diverse as meteorology, geology, and biology.

As many researchers would recognize by the 1980s, Lorenz's work also challenged the classical understanding of nature. The laws that Isaac Newton published in 1687 had suggested a tidily predictable mechanical system—the clockwork universe . . .

Unpredictability plays no role in the universe of Newton and Laplace; in a deterministic sequence, as Lorenz once wrote, "only one thing can happen next." All future events are determined by initial conditions. Yet Lorenz's own deterministic equations demonstrated how easily the dream of perfect knowledge founders in reality. That the tiny change in his simulation mattered so much showed, by extension, that the imprecision inherent in any human measurement could become magnified into wildly incorrect forecasts.[4]

Lorenz's discovery led to a big question: Why does a set of completely deterministic equations exhibit this behavior? After all, scientists have been taught that small initial agitations lead to small changes in behavior. This was clearly not the case in Lorenz's model of the weather. The answer lies in the nature of the equations; they were **nonlinear** equations. Nonlinear

---

[4]Dizikes, Peter, "When the Butterfly Effect Took Flight," *MIT News Magazine*, February 22, 2011.

systems and their mathematics are central to chaos theory and often exhibit fantastically complex and chaotic behavior.

In short, chaos embodies three important principles found in complex systems:

1. **Nonlinearity.** Complex systems are best described as web connections, not linear connections. They loop and feedback on themselves interactively. The degree of complexity resulting from dynamic interactions can reach an enormous level. Dynamic interactions are explained as high degrees of interdependencies, nonlinear interactions, short-range interactions, and positive and negative feedback loops of interactions.[5]
2. **Extreme sensitivity to initial conditions.** The effects of tiny initiating events, lots of them occurring in a short time frame, can produce significant nonlinear outcomes that may become extreme events. These are what John Holland calls "'small inexpensive inputs' or 'lever point phenomena'; my butterfly events."[6]
3. **Cause and effect are not proportional.** A part that costs ten cents can halt an assembly line as quickly as a $10,000 part.

Chaos and Complexity Theory Maturation

In 1984, the Santa Fe Institute (SFI) was founded by a group of distinguished scientists including Nobel Prize winner Murray Gell-Mann. Eight of the ten founders were scientists with Los Alamos National Laboratory. In conceiving of the Institute, the scientists sought a forum to conduct theoretical research outside the traditional disciplinary boundaries of academic departments and government agency science budgets. SFI's original mission was to disseminate the notion of a new interdisciplinary research area called complexity theory or simply complex systems. This new effort was intended to provide an alternative to the increasing specialization the founders observed in science by focusing on synthesis across disciplines. The intention was to overcome the "reductionist" thinking that had permeated science and academia and focus on discovering "the new rules" of underlying order of the universe that chaos theory implied. John Holland, a pioneer in genetic algorithms and adaptive computation, said

---

[5]Allen, Peter, Steve Maguire, Bill McKelvey, eds. "The SAGE Handbook of Complexity and Management," Sage Publications, 2011, page 123.
[6]Ibid.

of complexity science at the time, "At this point we are making guesses, feeling our way. But most of us think a science is there."[7]

In the 1990s, a group of scientists (including several Nobel winners) affiliated with the SFI decided, "There's not much point in studying chaos. It's too chaotic. Let's study complexity, where with the help of computers we can actually figure something out." They did—they found complex adaptive systems (CAS). Unlike *nonadaptive* complex systems, such as the weather, complex *adaptive* systems have the ability to internalize information, to learn, and to modify their behavior (evolve) as they adapt to changes in their environments. From this point forward, the science of complexity took off.

The science of complexity has to do with structure and order, especially in living systems such as social organizations, the development of the embryo, patterns of evolution, ecosystems, business and nonprofit organizations, and their interactions with the technological-economic environment. The following quote from Dr. Christopher Langton, a research scientist at the SFI and one of its founders, helps put complexity theory in perspective as to where it fits between Newton's ordered universe and the total randomness of chaos theory.

"Complexity" represents the middle area between order at one end and chaos at the other. Thus complexity is sometimes called the edge of chaos. If we think of order as ice and chaos as water vapor, complexity would be liquid water.[8]

Unlike ice which is a fixed solid, water has patterns, movement, and flow. It can be charted unlike the random movement of gas molecules.

Supply Chains as Complex Adaptive Systems

Proving the point that the shift in the view of supply chains from linear systems to complex adaptive systems (CASs) has made its way solidly into mainstream economic research is a 2008 paper, published by the *International Journal of Physical Distribution & Logistics Management*, the

---

[7]http://www.santafe.edu/about/history/
[8]Ibid.

authors, Wycisk, McKelvey, and Hu Eismann, cite over 79 references in support of their paper. Their abstract as well as their opening sentence in the introduction, confirms the "shift" is already here, that supply chains are indeed complex adaptive systems.

The important points of the paper are:

1. The idea that "butterfly levers" are the small initiating events that if identified and blocked, can stop large negative effects from occurring. Conversely, small triggers can also produce large positive effects.
2. The identification of the butterfly effect as the bullwhip effect associated with supply chains described in Chap. 3. *"As noted earlier, the CAS butterfly effect has already been observed in logistics systems and named the bullwhip effect. . . . It is typically observed in forecast-driven distribution channels.* **Like butterfly effects, the bullwhip effect describes how tiny initial shifts (in customer demand in order quantity) can result in chaotic and extreme events along the supply network via dynamic (nonlinear) processes. Owing to strong interdependencies among the actors of a supply network,** *the bullwhip effect takes place in human systems (logistics), where "intelligent" actors can worsen it through their conscious decisions of action and behavior."*[9]

Chapter 1 described one prevalent reason for the existence of the bullwhip effect—demand forecast updates in material requirements planning (MRP) and distribution requirements planning (DRP) systems. We believe there is another major reason for the bullwhip effect, and its effect is second only to the use of demand forecasts: The common use of a cost-centric efficiency strategy to determine policies, tactics, reporting information, and measures in the subsystems of supply chains. The resulting self-imposed negative variation spirals backward and forward into the supply chain and is the cause of large negative variations.

The Simplicity in Complexity

As the new kid on the block, complexity theory makes managing complexity "simple"—if you follow the right rules. In doing the research

[9]Wycisk, Mckelvey, and Hu Eismann, "Smart parts: supply networks as complex adaptive systems: analysis and implications, "*International Journal of Physical Distribution & Logistics Management,* Vol. 38, No. 2, 2008, pp. 108–109.

for this book, we were excited and amazed at the amount of writing and research that has appeared in the last eight years applying complexity theory to economics, supply chain, and logistics management. There has been a dramatic shift in the understanding of supply chains as logistic systems. It has evolved from linear structures to complex systems, and most recently to logistic systems being described best as complex adaptive systems. This change is a result of the applied sciences catching up to the natural sciences' shift from the Newtonian or a mechanistic and linear paradigm to include complexity theory and complex adaptive theory as an important part of the foundation for understanding natural order in the universe.

The acceptance of complexity theory alongside Newtonian theory has finally overflowed into the applied sciences. New research and journal publications are appearing with regularity. The new focus and research provides a solid foundation to use the breakthroughs in complex systems statistical models to rewrite the rules for management accounting and provide relevant information once again. Complexity theory validates constraint-based management principles and CAS is the foundation for demand driven approaches and smart metrics. It explains why companies using a demand driven strategy to develop new business rules, tools, and measures are making dramatic gains in ROI, DDP, and market share. CAS is the "new train" the applied sciences and the remaining management accountants have been waiting for.

It is important to point out that the new discoveries do not disprove Newton. What they revealed is Newtonian theory was insufficient *alone* to explain the universe, and not all phenomena were orderly, reducible, predictable, and/or determined. They set the stage for the acceptance that there was more to learn, and the emergence of chaos theory, complexity theory, and complex adaptive systems quickly followed. These theories offered the opportunity to explain and model both independent and dependent nonlinear systems. They set the context to be able to model, explain, and manage the complex, interdependent supply chains that are the reality of business in the twenty-first century.

Understanding Complex Systems

Today's supply chains don't look like chains anymore—they look like complex webs comprised of a significant number of nodes of manufacturers, transportation companies, and distributors. Flow of information and materials can loop and iterate in a nonlinear way. This means organizations must map the connections and flow of information (signals), energy, and materials between their subsystems to the extent of the boundaries of the system they want to manage.

> A systems view allows us to complement and add to the more common reductionist scientific paradigm, which focuses on one thing at a time, to the exclusion of everything else. Looking very closely at one thing can reveal important information. However, if we don't take the systems view into account, and instead depend exclusively on a narrow view, we are in danger of experiencing negative unintended consequences.[10]

Cases in point are the examples in Chap. 2, the Dreamliner case study in Chap. 9, "Company Normal's" story in Chap. 5, and Ford's failure to include the market in his systemic view in Chap. 8.

What Makes a System Complex?

A system is complex when the pattern of interactions between parts of a system creates a new level of order, and this new level of order results in systematic distances or delays between cause and effect in space or time.[11]

Today's global supply chains definitely fit the bill and the above sentence describes the bullwhip effect and companies' attempts to deal with it. The more complex systems are the greater the distance between causes and effects, which make them more difficult to track. Also, systems are dynamic, meaning they happen over time as they interact and move from state to state. Because systems are interconnected, when action is taken within the system, the system will have a response that can only be understood by understanding the connections and interconnectedness of the whole. One of the points of diagramming out the effect-cause-effect story of Company Normal is to demonstrate the interactions to different signals that move the organization from one state of being to the next

---

[10]http://www.csh.umn.edu/wsh/UnderstandingComplexSystems/index.htm
[11]Ibid.

system state. Signals have predictable effects both positive and negative, if the interconnections in the system are understood. This is why the first step is the creation of a demand driven operating model described in Chap. 3.

Complex Systems are Nonlinear

Complex systems are nonlinear and it is impossible to predict the behavior of the whole by analyzing the parts separately. This is in direct

Table 10.1 Differences between Newtonian and Complexity Theory

System Characteristics	Newtonian Linear	Complexity Nonlinear
The method to understand the system	Linear systems can be understood by studying the individual part; the whole is the sum of its parts	Nonlinear systems can only be understood by mapping and understanding the dependencies and interconnections of the parts
System predictability	The linear system state is stable and predictable	The nonlinear system state is dynamic, and no predictions remain valid for too long
System output behavior	The output of a linear system is proportional to its inputs	The output of a nonlinear system is governed by a few critical points—the lever point phenomena described above (Holland)
Mathematical models of the system	Gaussian statistical models (normal distribution curve)—The sum of the averages are a predictable model of the system; the tails of the statistical distribution can be ignored as anomalies	Paretian statistical models—The tails of the distribution contain the relevant information to predict and model nonlinear systems. The tails identify the few critical points and the relevant information to manage nonlinear system output—the lever point phenomena
System output maximization	A linear system can be optimized	A nonlinear system cannot be optimized but can continually improve

contrast to linear or mechanical systems, where the assembled parts work together in a prescribed way to provide a predictable function. An airplane or any other machine is not a complex system, even if it appears complicated. This is because no new levels of order will arise from an airplane or machine. It is bounded by the purpose and physics designed into it from outside of itself. It cannot change, learn, adapt, evolve, and emerge to a high order of performance. It can fly but it will never learn to be a submarine. It will never evolve beyond the limits set by its original design. It may or may not recover from a nosedive but that is dependent on the Newtonian physics involved in speed, distance to impact, and trajectory. It cannot move beyond, evolve, or change any of the physical laws limiting it.

The Move beyond Complexity Theory—Complex Adaptive Systems (CASs)

Complexity theory was the foundation of constraint-based management principles. Limiting factors are a reality but they are only one of the factors that must be considered when placing strategic control and decoupling points. Managing complex adaptive systems goes beyond identifying and exploiting constraints. It points to the underlying order in the universe that drives all complex adaptive systems to learn, self-regulate, adapt, and change. The characteristics of CAS behavior are the starting point to understand how to manage nonlinear systems as they truly have a mind of their own.

Figure 10.1 is a summary of important CAS characteristics that describe nonlinear system behavior. Smart feedback metrics are designed to complement and use the CAS laws of how people/the system will behave and the physical factors governing the system's output. The next section explains each characteristic, its behavior, and how to translate them to supply chain behavior. They are crucial to understand and consider when designing policies, tactics, reporting, metrics, and information technology tools. We will use the Company Normal example from Chap. 5 as our reference environment to identify the CAS behavior traits below and create a bridge to each of the CAS science terms. The behavior traits below were drawn from John H. Holland's book *Signals and Boundaries–Building Blocks for Complex Adaptive Systems* (The MIT Press, 2012).

CAS Characteristics	Definition and Examples
Boundaries	All "systems" boundaries are defined by their subsystems, hierarchies adaptive agents and **signal sets** to trigger actions and interactions between them. No "true boundary" exists in a CAS because they are always part of a larger whole. The Demand Driven Design process models boundaries at the limits of the adaptive agents ability to "act".
Coherence	All CAS depend on the subsystem to align its purpose to the purpose of the greater system. Subsystems that act in ways that endanger the greater system it depends on lack **"coherence"**. A lack of coherence can push the system over the "edge" and into "chaos".
The Edge of Chaos	All CAS operate in the transition zone between stable equilibrium points and complete randomness - chaos. These zones have been "learned" and are triggered by **"signals"** recognized by the system's adaptive agents. They will mutually self limit their actions to avoid crossing over into "chaos".
Self-Organization, Innovation and Emergence	All CAS have **"emergent properties"**. Adaptive agents will **self organize** to solve an issue when it becomes visible and triggers a signal they recognize requires new action. Through learning over time, their new interactions will emerge the system to a higher order and system **innovation "emerges"**.
Signals	All CAS use defined signals to communicate inside their subsystem and with subsystems whose boundaries interact with theirs. **System coherence and signal alignment is a primary consideration in determining signals, their triggers and instruction sets**.
Adaptive Agents	All CAS are dependent on **adaptive agents** to intercept, understand and pass on the signals they receive. They are adaptive because they can learn and recognize new patterns to add to their signal set of "if...then" conditions to trigger actions.
Signal Strength	All CAS depend on the adaptive agents to recognize when two signal actions are competing for the same time step or resource and prioritize the signals based on the perception of the strength of the signal to align to the "system goal". **Signal strength should be based on keeping system coherence.**
Self Balancing Feedback Loops	All CAS adaptive agents depend on **feedback loops** to understand "the state" of their subsystem to trigger their signal "instruction set" and determine the signals strength and priority. Adaptive agents use the feedback loop to provide relevant information for receiving, interpreting and passing on signals.
Resilience and Rigidity	Resilience is how well a system can return to stability when it experiences random or self-imposed variation. Rigidity is the opposite of resilience. If a system has become rigid, then it is fragile and more apt to slide into chaos..

Figure 10.1 A Summary of Complex Adaptive System Characteristics that Describe Nonlinear System Behavior

Boundaries

Sometimes we can only see the system by looking at the connections between the parts of the system. There is no obvious physical membrane that lets us know where the boundary is in a supply chain. An example:

> There is a lathe in the machining center, in a manufacturing plant, in the operating division of applied friction devices (brakes) that work on parts for the braking system for automobiles, heavy trucks, airplanes, and heavy equipment; supplying both new vehicle manufacturers and replacement parts for existing vehicles already in service in different fleets, serving different industries around the globe. Some of the parts it produces are make-to-stock and some are make-to-order. A demand signal necessitating the lathe to work can come from any of the myriad of global market niches in the supply chain, but the lathe

can only begin work if the flow of signals has brought the right materials, tooling, supplies, and instruction set along with a signal to begin the work, to the lathe operator.

The point is that no boundary is absolute because everything is connected. There is never any division by which one can say that one thing is absolutely a part of a system and something else is absolutely not part of that system. But we can make a simple, practical map of the system's flow, the signal points, and the boundaries between the subsystems and the system we are going to manage.

Coherence

A complex adaptive system's success depends on coherence of all of its parts. A subsystem's purpose has to be in alignment with the purpose of the greater system in order for there to be coherence. Without that alignment, the subsystem acts in a way that endangers the greater system it depends on. Coherence must be at the forefront of determining the signal set, triggers, and action priorities. To keep coherent, adaptive agents must ensure that their signal sets contain the relevant information to direct their actions, and that they are not at cross-purposes with the goals of the systems it depends on.

The Edge of Chaos

Yet these adaptive agents, with conflicting goals, will still coalesce and work together even when it means limiting or compromising their own subsystem's goal. Why? The cliché, "politics makes strange bedfellows," exists for a reason. These collaborative compromises between the adaptive agents oscillate the company between mutually-agreed-to boundaries because the agents jointly recognize that crossing over the edge of chaos puts the system and all of them at risk. Crossing the edge into chaos will force a change that managers know they cannot manage and may not survive.

Mutual, even collaborative, self-limiting behavior of managers in supply chains points to three important conclusions:

1. The subsystems signal sets, measures, and goals lack coherence to the system goal they operate in.

2. Suboptimal system performance is guaranteed and is the major source of the oscillation effect and the resulting negative self-imposed variation explained in earlier chapters.

3. Innovation is often stifled. The system stays in an oscillating pattern moving them from one suboptimal system state to another suboptimal system state. This is perceived as the only safe way to operate. This safe, but far from optimal, zone fits the limit of the adaptive agents knowledge of how to manage their system with their existing rules, tools, information, and measures that make up their feedback loop. Innovators are punished or generally discouraged because they endanger this safe zone.

Self-Organization and Emergence

A complex adaptive system will self-organize without any outside force directing that organization. Purely through parts of the system interacting, a more complex pattern emerges. The concept that all CAS have emergent properties that arise spontaneously as the system self-organizes is key to understanding the importance of giving people relevant information. The term *emergence* refers to this spontaneous arising of new levels of coherent order. Emergence and self-organization go hand in hand. People will self-organize around issues they perceive need to be solved as the issues are made visible. Providing visible, coherent signals based on relevant information, keeps the subsystem goals aligned with the system's goals.

Coherence allows the system to continually emerge to a higher state of performance—continually improve. Coherence can only exist if the cost behavior and flow rules match the behavior of the system rules to which they belong. Adaptive agents will coalesce with the other resources necessary to take the action to keep the system out of chaos. For organizations to take advantage of this behavior and continually improve, the signal sets must be visible, real-time, and based on relevant information, coherent to the system's goals.

People will self-organize to solve problems. If the observed problem is the result of a signal or measure at odds with the system's goals, then the problem will never be resolved no matter how much money and energy is thrown at it. It is merely a symptom, an effect of a coherence core problem. They will repeatedly solve the same problem over and over again as the system oscillates between repeated and competing course corrections.

These incoherent signals or measure, imposed on a system, block a complex adaptive system from learning and improving. People in such a system come to view this behavior as normal and expected. It becomes part of their paradigm of how the world works—their deep truth.

Emergence and Innovation

The self-regulating feedback loops and signal sets of the adaptive system agents control the tendencies of the system to go too far to the extremes and prompt actions to keep the system in check at the edge of but not into chaos. The agent uses this space to learn and adapt from the interactions with the system outside of its subsystem's boundaries. Through learning, over time, a new level of order emerges from the interactions and provides the next higher level of order for the system to operate from. Once again, this is known as emergence. All complex adaptive systems have emergent properties. Complex adaptive systems push to the edge of chaos, learn and evolve, or decay.

John Holland expressed it well:

> Innovation is a regular feature of complex adaptive systems. Equilibrium is rare and temporary. Adaptation by recombination is a continuing process at all levels. . . . Any time a new kind of agent appears, there are multiple opportunities for new interactions that modify temporary local equilibrium. The new interactions make possible further interactions and adaptations because the agents incorporate adaptive mechanisms, the systems continue to innovate. In other words, the mechanisms of change are primarily mechanisms of exploration rather than exploitation.[12]

The emergence of the knowledge transfer of work from skilled craftsmen, to precision machinery and tools that allowed for the innovation of mass production is one example. Supply chains are continually adapting and emerging to higher levels of order from their interactions with the larger system of which they are part. Ford's innovations on Taylor's work took

---

[12]Holland, John H., *Signals and Boundaries—Building Blocks for Complex Adaptive Systems,* Massachusetts Institute of Technology, 2012, page 59.

supply chains to a higher level of order—the ability to mass produce a complex product. Donaldson Brown and Sloan's innovation on Ford's work took supply chains to a next higher level of order—a proliferation of complex products segmenting markets that included financing and marketing innovations by the new agent, General Motors. Ford's failure to sense and adapt to the appearance of a new kind of agent resulted in the closing of the Model T factory in spite of all of his previous innovation. Today's complex and connected global economy has no shortage of new agents.

Self-Balancing Feedback Loops Maintain Stability

In a nonlinear system, feedback loops occur when things affect each other in a circular path. Think of a thermostat as an example. Balancing feedback allows a system to self-regulate, so the tendencies of the system to become extreme are held in check. Adaptive agents have learned to not cross over the edge of chaos. Perhaps they have "been there, done that." Providing visibility to a balancing feedback signal prompts a complex system to change behaviors and to adapt more skillfully to changing conditions. Feedback loops frequently have delays, which make tracking the feedback more difficult. The greater the feedback delay, the more oscillation and waste.

Resilience and Rigidity

Resilience allows a system to respond to a disturbance while maintaining equilibrium within its system boundary. In supply chain words, resilience is how well a system can return to stability when it experiences random or self-imposed variation. Resilience arises from the subsystems' abilities to respond to the feedback loops that regulate equilibrium. The more flexible the resources are, the better ability to adapt and repair the pattern. The ability to adapt and the diversity or flexibility of options/actions determine how quickly the system can recover and or improve. This is what makes protective capacity buffers at noncritical resources so valuable. The more they can flex to recover and or rotate to areas in trouble, the more resilient the system is. Conversely, if the subsystem lacks goal congruence, this flexibility speeds up the oscillation effect building and passing on variation through the supply chain.

One of the key lessons from the story of Henry Ford is that he made his logistical system a bureaucracy that was extremely efficient but too rigid or too narrowly focused. He did not include the market in his signal boundary and missed the obvious market and consumer changes. He missed the interconnectedness and importance of both financing and the used-car market on his ability to market and sell the cars he produced. He ignored the change in the consumers' view of value and his competitors' response. His narrow definition of the market caused him to miss market opportunities including the opportunity to segment his market without segmenting his resources, a strategy General Motors developed and exploited.

The opposite of resiliency is rigidity. If a system has become rigid, then it is fragile. An example is the lean distribution systems that move goods around today. It can be very efficient, but it is also fragile. If it falls apart, most cities only have a few days of food before the shelves will be bare. Leaning out too much can create a brittle system that cannot easily recover when the unexpected occurs in a supply chain.

A Complex Adaptive System Example—Company Normal

Company Normal is our reference environment to identify adaptive agents, signals, signal strength, the edge of chaos, self-organization, and emergence in a supply chain environment. The plant manager is the adaptive agent in charge of his subsystem and the adaptive agents he manages are the demand-and-supply planners, production schedulers, production resource managers, and inventory stock managers. The signals are purchase orders and work orders, stock status, resource schedules, resource loads, dispatches lists, and expedite lists. Operations interact with the adaptive agents from Finance, Sales, Engineering, and Quality as well as other subsystems whose boundaries interact and send or receive signals inside the Company Normal system. These managers interact at specific time frames triggered by financial reporting measurement periods for monthly and quarterly meeting reviews. They will also interact anytime the system feedback loop triggers a warning that one of the managers can no longer ignore. For example, the loss of a major sales account is an example of a system state that is new information to the plant manager, requiring him to learn more and adapt—perhaps even change his instruction set if warranted. In the Company Normal story from Chap. 5, the impromptu conversation between the sales manager and the plant manager demonstrates this point.

> Walking out of the room the plant manager asks the sales manager what accounts they have lost and who they are in danger of losing next. It now becomes clear to the plant manager that he is going to take some heat from at least one of the vice presidents, and maybe even the CEO, and very soon. Now he begins to worry. He tells the sales manager he understands and will take some actions when he returns to the plant to crank up the pressure on those orders even if he has to break into set-ups, set other orders aside, and rob the raw materials from other jobs. They will work Sundays if necessary. The sales manager thanks him and both hurry out to make phone calls to change priorities.

The trigger point for the plant manager to change signal strength and instruction sets is dependent on his perception of the edge of chaos. His indication that due date performance is not good enough is a direct market signal, not the tracked DDP or calculated stocking service level percentage. This new information increases the plant manager's perception of the level of threat and increases the signal strength of the sales manager's concerns and request for tactical changes. The plant manager processes the information, makes a decision to act, and sends out a new instruction set changing priorities and actions that will ripple out through the supply chain. The actions and priorities he set in motion will create the system state for next month's meeting. The fact that he is willing to self-limit his subsystem's primary performance goal is an indication that market service has dropped too low—he now must compromise. This recognition forces him to shift his signal strength priority from focusing on his own primary goals to focusing on Sales' primary goals and their instruction set.

At this point they simply do not have a better method. It has worked well enough in the past to keep them out of chaos and keep the plant manager employed. We may also conclude that their competition must not be doing any better. The market has not forced them to innovate or face extinction. Company Normal fits perfectly with the behavior traits of complex adaptive systems listed above.

Reference Examples for Complex Adaptive Systems

Adaptive agents in all supply chains should act like air traffic controllers for the subsystems they serve and the system of which they are a part. The air traffic control system is a good example of a complex adaptive system where adaptive agents function to manage flow with the goal of keeping the subsystem coherent to the system goal—safe and on-time delivery of airplanes. To do their job successfully, they must have a constant visual and audio update of the status of the dynamic state of traffic in their airspace boundary. They receive and give warnings as well as direction for changes when necessary. Every airplane has a scheduled departure time, a routing, and a scheduled landing time. Departures and arrivals are carefully staggered to allow for an orderly flow at the rate their originating and receiving runways can process them.

Airports with heavy traffic loads have a strategic buffer of airplanes on the ground waiting to take off in the right order and spacing as well as a schedule of airplanes staggered to land. If the airport experiences enough negative variation, there will be a queue of airplanes in designated holding patterns, waiting for their signal to land. These buffers are carefully sized, managed, and controlled based on the system state. The controllers will send and receive signals to the airplanes and the airports based on the appropriate instructions for the current system state. A decision to choke or delay departures often occurs if destination airports experience enough negative variation to create an overloaded buffer of airplanes in holding patterns at the destination airport.

The controllers are in direct communication with pilots and each other where and when their boundaries meet, to ensure coherent signal transfers from one airspace boundary to the next. They determine priority to land and take off on their definition of signal strength based on the airplanes' scheduled arrival. These scheduled arrivals directly influence if the plane and air crew can make their next scheduled departure and keep coherence for the whole system of air traffic. Airplanes with emergency conditions jump signal strength to the front of the schedule priority and are expedited into the landing queue based on the level and type of emergency. This reflects their signal set coherence with the system goal of safety. Airplanes are rerouted based on the airport delay time frame and their available fuel range. Their signal set contains triggers associated with a "defined system state" of when they must act and the instruction set for the potential actions to take, as well as a list of resources to send signals to when the system state's feedback loop measures and recognizes the trigger point has occurred. Time lags between the actual system state and the visibility to the adaptive agent and their capacity to act add variation and risk. Their system signals are kept as real-time as possible.

CAS are dependent on their adaptive agents' ability to intercept, understand, and pass on the signals they receive in order to keep the goal of the system coherent. They are adaptive because they can learn, recognize pattern changes, and add to their conditional statements of "if . . . then" connections to include the new potential dynamic system states that require developing new action/messages to pass on. The agent's job is to keep the system coherent to its goal. If the signal set's rules are flawed, the actions taken will not produce the new system state predicted and expected. This will cause the system to experience variation, delay, and disruption and require additional action or sets of action to correct. The faster the adaptive agents learn and change the flawed assumption about the rules governing their system's interactions and behavior, the more successful the system will be. It improves and emerges to a higher state of order.

In both nature and human systems, decay and eventual extinction is the result when adaptive agents and their signal sets are too far from coherence with the larger system they exist in. It is this awareness that causes adaptive agents to work together and mutually self-limit to keep themselves and the system safe from the edge of chaos.

In Company Normal, the subsystem boundaries are clearly drawn between operations, finance, sales, and engineering. All have their own adaptive agents, signal instruction sets, and signal strengths, based on meeting their own goals.

▲ Operations' signal set is focused on meeting planned production output, minimizing unit cost, and maximizing resource efficiencies and production overhead variances.

▲ Sales' signal set is focused on clearing the sales backlog of past-due orders, stock-outs, and increasing DDP.

▲ Accounting is watching all of the KPIs but is currently focused on the skyrocketing inventory and the low cashflow.

All three sub-goals have different signal sets and different tactical instructions. When a tactical priority change is triggered, the actions and priorities of the planners, buyers, schedulers, and resource managers change. The instruction set the plant manager follows and passes on to his resources is based on his interaction with all of the subsystems' managers. They all know the triggers that indicate the system state has approached the edge of chaos. They know they must change their behavior and attempt to rebalance the system. Ignoring the imbalance can push the subsystem over the edge of chaos, threaten the system's ability to perform, and risk their ability to stay employed. Companies continually ride this teeter-totter staying in some established or benchmarked KPI or cash-flow boundary. But not always . . . Failed companies who passed over the edge of chaos and did not maintain sustainable cash flow to recover are not uncommon. Successful adaptive agents have learned when to switch signal priority and signal instruction sets when a KPI falls or rises to the danger zone.

In Company Normal, all three competing signal instruction sets result in different suboptimal system states:

1. Unacceptable level of service
2. Unacceptable inventory performance
3. Unacceptable gross margin (price discounts, unit cost, resource efficiency)

The time spent in each state and the relative trade-offs between them are determined by the interaction between the adaptive agents and the three different signal instruction sets. The system state they are currently oscillating towards or away from is determined by the adaptive agents taking actions based on their perception of signal strength determined by the KPI in the most trouble at any particular time.

Figure 10.2 depicts how these three factors relate to the edge of chaos for an organization like Company Normal. The figure is divided into three pie slices: inventory, service, and gross margin. The dark outer ring represents insolvency—the definition of chaos for most organizations. Any of the three factors can push the organization into insolvency by either consuming cash or not providing enough cash. As an organization's position within any of the pie slices moves closer to that outer ring, signal strength goes up. The center circle represents equilibrium and coherence to the system goal between inventory, service, and gross margin.

Company Normal's feedback loop exists to keep the system from going too far off the edge of unacceptable service levels at one end and unacceptable inventory investment and standard unit cost and margin

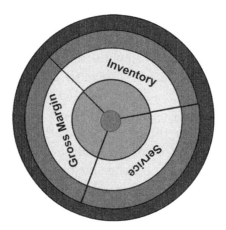

Figure 10.2   Signal Strength and the Edge of Chaos

performance at the other ends. Managers' actions oscillate inside some bounded realm of sustainable cashflow or solvency. Regardless of the current system state's position, both the inventory oscillations and service-level oscillations compromise the system objectives of high DDP and high ROI. The oscillating actions generate expedite-related expense, bloat inventory, erode service levels, and deplete cash. Passing the boundary jeopardizes the system goal or planned ROI and jeopardizes the adaptive agents' ability to survive. No cash or the inability to obtain more cash defines the limits of an organization's ability to survive. The key point is Company Normal is stuck inside a bounded range of limited performance. Because of the subsystem's lack of relevant and timely information, coupled with local KPIs that lack system coherence they cannot learn and emerge to a higher level of performance.

Connections and Interactions in Nonlinear Systems

Wherever connections happen, patterns emerge. Understanding these connections makes the complex system patterns predictable as a network of cause-and-effect loops. The sheer mass of data in a network can hide the common patterns of the flow of signals to and from resources. The non-linearity of complex systems means that despite a high level of seemingly complex connections and interactions, there can be an inherent level of simplicity in controlling them. Some points in the system are more critical than others in determining the patterns and the outcomes. These critical points set both the definition of relevant information and the limits of the relevant time frame to define the economic short run. The short run is the timeframe in which the cost, revenue, and output volumes will remain stable. Identifying these few control points and decoupling points gives an organization the ability to generate a relevant cost-volume-profit analysis for the time period and provide relevant information for tactical decisions and actions tied directly to ROI impact.

In a demand driven operating model, the control points and decoupling points are the butterfly levers or the lever point phenomena of nonlinear systems. One of the three important principles of complex systems previously mentioned is their extreme sensitivity to initial conditions. The effects of tiny initiating events, lots of them occurring in a short time frame, can produce significant nonlinear outcomes that may become extreme events. By creating a visible feedback loop focused on the state of the control points and decoupling points, these tiny initiating events can be identified and blocked from being passed on and spiraling into large negative effects in the system.

The demand driven operating model, explained in Chap. 3, is the starting point to managing flow in nonlinear systems. A short review is appropriate to show how it uses the CAS behaviors discussed above. Designing a demand driven operating model is the positioning part of position and pull. To get the positioning right, two things are required:

▲ The identification and placement of decoupling and control points
▲ The consideration of how to protect those decoupling and control points

Placing Decoupling Points. Decoupling points are places to disconnect the events happening on one side from being directly connected to the events happening on the other side. They delineate the boundaries of at least two independently planned and managed horizons. They break the accumulation of variation, tiny initiating events (TIEs), from being passed forward and backwards. These decoupling points are most commonly associated with stock positions, but all three strategic buffer types (stock, time, and capacity) block and break variation.

Placing Control Points. Control points should be thought of as places to transfer, impose, and amplify control through the system. Control points are often placed between decoupling points with the objective of better control in the lead-time zones between those decoupling points. A shorter lead time and less variability associated with that lead time results in less required stock at the decoupling point (a working capital reduction) and a smaller time buffer (a lead-time reduction).

> *Consideration 1: Points of Scarce Capacity.* The total system output potential is directly limited by the slowest resource.
>
> *Consideration 2: Exit and Entry Points.* When work leaves our system it leaves our effective control. Carefully controlling that entry and exit allows us to know whether delays and gains are generated within our system or outside of it.
>
> *Consideration 3: Common Points.* Common points are points where product structures or manufacturing routings either come together (converge) or deviate (diverge). One place controls many things. "All roads lead to or come from Rome."
>
> *Consideration 4: Points that have Notorious Process Instability.* When a resource is wildly out of steady state, making it a control point forces the organization to get it under control.

CAS and the Demand Driven Operating Model

Identifying the system decoupling and control points and understanding their behavior makes complex systems simple and predictable in the short run. The short run is the time frame in which the assumptions the model is built on will remain valid. Pareto distributions best explain nonlinear systems because they model the large effect of the very few factors that govern the whole system. Pareto distributions are used to identify places to decouple and control. Small positive or negative changes in them have a large impact on the system; they are the butterfly levers. The lever points will identify themselves when the connections are mapped and the system patterns emerge using the four considerations above for control point placement. These critical points (control points and decoupling points) identify the minimum necessary relevant data to model, understand, and control the system. The control points and decoupling points determine the relevant time range to plan, execute, and measure successfully. We can practically and safely exclude the mass of all other data as irrelevant and make simple, realistic prediction models for reliable planning, scheduling, and execution.

Data Collection and Feedback Loops

You can't measure what you can't see. Data collection for smart metrics is designed around decoupling and control points and the buffers that protect them. They provide four important visibilities:

1. Visibility to the current state of both decoupling and control points for planning, scheduling, and execution decisions
2. Visibility to the status of the strategic buffers protecting decoupling and control points from disruption
3. Visibility to the status of a particular order in relation to control points
4. Visibility to tactics for emergence. Monitoring the time buffers zone penetrations exposes and allows issues to be resolved proactively. This protects both the critical points and the system's schedule stability and reliable output performance to plan. The events occurring in the tails of the distribution of the buffer points are the lever point phenomena.

Feedback loops with delays make tracking and managing more difficult. The timeliness of the right signals about the right things is extremely important. Remember, all benefits will be proportionate to the speed of relevant materials *and* information. Constant feedback loops centered on the critical points and particularly the buffers that protect them allow the system to dial itself into the demand driven operating model for maximum responsiveness with minimum investment and waste.

If you turn on the hot water for a shower, it may take a while for the hot water to get from the hot water heater to the shower. In the meantime, you've cranked up the hot. When the hot water finally arrives and is too hot, you crank on the cold, which doesn't seem to do anything right away, so you crank it further. When the cold arrives, it is too much. This oscillation goes on until you close in on the desired temperature. It wastes water and energy, and takes longer to get through the morning shower. Alternatively, we could install a hot-water circulating pump on a timer set for the hours we shower and eliminate the delay. We could also buy a shower faucet that can be preset to deliver the right water temperature every time we turn it on. This is the concept behind strategic buffers.

Subsystem feedback loops need to be as real-time as the system requires to stay on track and coherent to the system goal inside. The frequency of data updates is a function of the practical cycle time of the subsystem—no more and no less. If planners only release/process work orders once a day (frequency of sending a signal) then refreshing the state of incoming supply and demand signals every minute provides no value. If orders are processed continually throughout the day and pulled from stock to ship continually, then refreshing supply-and-demand signals in short intervals has value because it stops the variation and time delays associated with batching or latency. If execution cycle times are short, real-time updates are critical to avoid delay in synchronization.

Delays in supply chain signals almost always result in an inventory investment to protect the next link in the supply chain over the increased time from the delay. The greater the delay, the higher the variation associated with that delay. The higher the variation, the greater the chance that inventory investment is misaligned with the actual requirements. This creates the classic case of too much of the wrong and too little of the right. Once again, the value of Plossl's first law introduced in Chap. 1 can be seen.

Demand Driven and Resiliency versus Rigidity

A demand driven strategy uses three types of strategic buffers—capacity, time, and stock to create reliability, stability, and resiliency in a system. Monitoring the state of these strategic buffers creates visibility to the control point's ability to stay on schedule. This enables action to prioritize the right work and an early warning system to take action to correct work whose schedule is in trouble. Strategic buffers have five core functions to manage and measure flow.

1. They ensure strategic stored time to recover from variation and protect flow at control points and decoupling points and make a system both resilient and stable.
2. They provide a visible feedback loop on the state of the system through a view of the state of the critical points (control points and decoupling points) the buffer is designed to protect. Buffers are designed to flex inside pre-determined ranges delineating "too much" and "too little" time protection.
3. Visible strategic buffers that are correctly sized, monitored, and measured provide the signal, signal alignment, and signal strength. Signal strength to prioritize work when resource contention (either materials or resource capacity) exists is determined by the percentage of buffer erosion. The greater the erosion, the higher the signal strength

and priority. The concepts of visibility to the state of strategic buffers and the control points are the fundamental building blocks of relevant information. They are the feedback loop the system uses to measure and provide direction to self-balance. Smart metrics evolve from them.

4. Buffer status highlights emerging problems proactively for people to self-organize around and solve. This is true for both short-run planning and execution as well as Pareto analysis over time to track emerging patterns and focus improvement efforts and investment spending.

5. Buffer trends over time capture the patterns of the system. Better understanding the patterns leads to better planning and scheduling. The more realistic the plan, the more reliable execution. A shift in a pattern can direct time-standard corrections, or point out when a change to either the placement or the sizing of strategic buffers is warranted due to a product or market shift. Improvement gains that decrease variation, increase velocity, and free up both capacity and invested capital stored in inventory protection. This cycle of improvement gains allow strategic buffers to safely shrink and create a positive loop in ROI.

Determining whether a system is resilient or rigid involves monitoring the system over time. Systems have behavior that occurs over time and the patterns can only be discerned over a period of time. Therefore, static snapshots can't give you the whole picture. As materials or information flow into a system, it is continually altered. This is why all feedback loops such as strategic buffers, must have two capture points relative to time.

▲ The first provides monitoring and metrics of the current state of the system's critical points (control points and decoupling points). It is used to align priorities and focus action for planning and execution priorities to focus flow to the system's purpose. This view should be in real-time as the state of the system is dynamic and moves from system state to system state because resources are constantly taking actions.

▲ The second is to capture trend patterns over time of the system anomalies—the tails of the distribution of what disrupted or endangered the critical system points can be captured through tracking the events penetrating tail zones of the buffers that protect the links and linkage. More on this in the next chapter, but these disruptions can be captured and categorized by root cause to direct process improvement and or investment with the highest benefit to enhance or improve the system flow.

The demand driven operating model focuses on maximizing flow and velocity between strategic time and stock buffers. Strategic buffers by definition are time stored and strategically placed to pool either work-in-process or materials, subassemblies, and/or finished goods in order to break and absorb dependent variation. They block variation from disrupting the critical points (control points and decoupling points) that govern system flow and create schedule reliability and stability. If accumulated variation collapses a strategic buffer, the speed to rebuild them is a measure of the system's resilience. It is the reason for planned protective capacity. Balanced lines are rigid and can be brittle and subject to disruption when variation occurs. The more variation a system experiences, the more protective capacity is required in the form of strategic buffers to protect flow. Thus, strategic buffers *are not waste;* they are *required* to protect flow/velocity.

Summary

Supply chains are complex adaptive systems. The success of a supply chain strategy is dependent on signal alignment to speed relevant flow of information, materials, and resources to and through the subsystem boundaries. Coherence is only possible if the signals pass on only relevant information, materials, and resources. This is possible only with a visible feedback loop that proactively identifies and prioritizes critical issues for resolution as they arise. Visibility across the supply chain to the state of critical points (control points and decoupling points) is necessary for people to self-organize and the supply chain to emerge to a higher order, that is, to improve itself. Reactive management is a disaster in a complex supply chain.

The authors are proposing a new cause for the bullwhip effect, which is equally as significant as forecast error—the self-imposed variation created from the use of irrelevant information driving execution and investment decisions—"the bad measures". The bullwhip effect can be tamed only when both are addressed successfully. The variation from the use of forecast demand and a cost-centric efficiency strategy and its policies, tactics, reporting, and measures desynchronizes coherent flow signals, drives competing actions, and destabilizes the whole system.

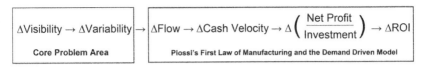

Figure 10.3 The Area of Deeper Truth versus Plossl's First Law

We can quantify the performance gap and the opportunity between the "old rules" and the "new rules."

The formula in Fig. 10.3 is how we define the performance gap mathematically in demand driven strategy terms:

▲ Visibility is defined as relevant information for decision making.
▲ Variation is defined as the summation of all the differences between what we plan to have happen and the effects of all the small input course corrections that cumulatively spiral into the bullwhip effect in supply chains and the butterfly effect in nature.
▲ Flow is the rate in which a system converts material to product required by a customer.
▲ Cash velocity is the rate of net cash generation (sales dollar – truly variable costs) – operating expense.
▲ Net profit/investment is, of course, the equation for ROI.

The equation in Fig. 10.3 is based on the CAS behaviors and assumes people will act on the information and signals they are provided and according to their metrics.

If visibility to relevant information improves, variation is either negated or removed, velocity of flow increases, and ROI opportunity increases. Conversely, if visibility to relevant information decreases, variation increases, and the inventory, labor, and/or capital must increase to compensate for the increase in the variation, or velocity and output will go down: the ROI opportunity will decrease.

Conclusion

The hard sciences struggled greatly to overcome their own attachment to Newtonian physics as the umbrella capable of explaining the sum total of the universe. It is no surprise there is a dramatic performance gap, a void,

214 | Chapter 10   Complexity Science and Supply Chains

between some of the applied sciences and a chasm between the current business rules and tools managers use today and the new business rules and tools they can/should use to manage their complex supply chains. Under the new paradigm of complexity theory and the emerging theories on complex adaptive systems, the old strategy and rules based on the reductionist view of individual resource or unit efficiency and cost fall apart; case in point: the Boeing Dreamliner presented in Chap. 9.

Although complexity theory is new as a paradigm applied to business management, complexity science has revolutionized how the hard science sees the world. Complexity theory and the constraint-based management principles derived from it, coupled with complex adaptive system behavior rules are the foundation used to define a demand driven strategy, policies, tactics, and measures. They are grounded in the hard sciences and reflect how the world works not how we think it should work or how we would like it to work. The greater the distortion between our model of reality and reality itself, the greater the variance between the result we expect and the result we actually get. The first step with any system or supply chain is always to map the link and linkages and cause-and-effect of the system.

In Chap. 11, we will take you from a beginning-to-end case study demonstrating how a complex supply chain can practically apply all of the elements we have discussed and create smart metrics that drive continuous system improvement.

CHAPTER 11

Smart Metrics

Managing in a Pareto World Calls for New Thinking

The main problem we are dealing with is *not* a lack of known methodology, rules, information tools, and smart metric feedback loops. Companies in a wide array of industries and differing product complexities have successfully implemented and sustained demand driven performance systems in some cases for over a decade. The problem is ontological: what type of reality do we assume we are attempting to control and manage?

The reality of an organization's supply chain processes and interdependencies determines its logistical realities and dictates its market and investment strategy. This reality defines the rules used to determine relevant information and how cost, revenue, and flow behave. A balanced supply chain performance measurement system has three distinct components of financial and nonfinancial metrics; one and two are the subject of this chapter:

1. Internal financial measures for evaluating strategic investment decisions. They follow the relevant information rules for nonlinear systems previously discussed. Those rules determine the behavior of the costs, revenues, and volumes over time. The starting point is a demand driven system design to agree on the market lead-time strategy and determine the strategic investment required to deliver the agreed-to system strategy.

2. Nonfinancial measures (day-to-day control of manufacturing and distribution operations) for control point and decoupling point buffers, feedback loop systems, and smart metrics are used to measure a supply

chain's performance. The primary goal is system coherence and signal synchronization to the defined strategy.

3. GAAP period external reporting has been previously discussed and is not relevant to the subject matter of this chapter.

The point of the previous chapters is to create a sufficient case to challenge the paradigm of supply chain performance management as independent-event financial data points managed and measured through an additive series of static snapshots of GAAP unit costs. Financial measures have no place in the day-to-day supply chain management of manufacturing and distribution. This chapter assumes the shift in paradigm has been made and offers a different view of the problem and a different solution set, metrics, and improvement opportunities. A Paretian view focuses on a visible feedback loop of strategic buffers of stock, time, and capacity and the control points they protect. Together they create system coherence and a set of measures focused on the six smart metric objectives.

Figure 11.1 is a summary of the six measurement objectives of a demand driven performance measurement system. The first four are the nonfinancial day-to-day operations measurement objectives of smart metrics:

1. Reliability
2. Stability
3. Speed/Velocity
4. System Improvement and Waste Opportunity

		The 6 Metric Objectives	Measurement Objective Definition and Examples
Non-financial measures	Day to Day Operations Control	Reliability	**Reliability** – Consistent execution to the plan/schedule/market expectation;
		Stability	**Stability** – Pass on as little variation as possible;
		Speed/Velocity	**Speed/Velocity** – Pass the right work on as fast as possible;
Both	Strategic Decisions	System Improvement & Waste (Opportunity $)	**System Improvement/Waste (Opportunity $)** – Point out and prioritize lost ROI opportunities.
Financial		Local Operating Expense	**Local Operating Expense** - What is the minimum spend that captures the above opportunity?
		Strategic Contribution	**Strategic Contribution** –Maximize throughput dollar rate and throughput volume according to relevant factors;

Figure 11.1 Summary of the Six Metric Objectives for a Demand Driven Performance System

There are three financial strategic measurement objectives (notice that list item 4, System Improvement, is both strategic and day-to-day), as well as a mix of financial and nonfinancial measures. A systemic strategic investment analysis requires a mixture of financial measures but not GAAP rule-based financial numbers or measures:

4. System Improvement and Waste Opportunity
5. Operating Expense
6. Strategic Contribution

The Power of Pareto and Strategic Buffers

The most important thing for managers to manage is the events outside the targeted limits. In particular, a shift of managerial attention from the center of the distribution (the averages) to the tails (the outliers) reveals solutions to existing problems and promising opportunities for both market growth and process improvement. Smart metrics focus on the strategic control points and decoupling points. The events occurring in the tails of the strategic buffer zones trigger management to act. Purchasing, planning, scheduling, and deployment decisions are determined and aligned, and execution is synchronized by buffer zone priorities. Events in the tails are measured and trended to determine resource and asset performance as well as focus improvement opportunities.

The authors make the following assumptions regarding all of the examples and the example company case used in this chapter:

▲ The organization has embraced a strategy of flow-centric efficiency.
▲ A demand driven design strategy has been completed and decoupling and control points have been chosen to protect the agreed-to strategic market lead time.
▲ Demand driven MRP (DDMRP) methodology has been used to determine stock strategy at the decoupling points.
▲ The stock, time, and capacity buffer zones have been initially sized correctly to protect flow with the minimum investment.
▲ The organization has created the ability to visibly display these buffers in real-time status.

DDMRP is not the subject of this book, although all of our assumptions regarding inventory management and supply and demand planning are based on the principles of DDMRP as defined in *Orlicky's Material Requirements Planning*, third edition, by Ptak and Smith. A short summary on strategic stock buffer zone calculations is provided in Appendix A. For a detailed understanding of both the math and principles of DDMRP and strategic stock buffers, the authors recommend the above book. For our purposes, we will assume the math behind the zones in our examples is correct.

Stock Buffers, Pareto Analysis, and Smart Metric Objectives

Figures 11.2 through 11.4 demonstrate a Paretian view of stock buffers and the smart metric system objectives. Figure 11.2 is a spectrum view that exists with all inventories either at the single item or aggregate level. In Fig. 11.2 you see a line running in both directions. This line represents the quantity of inventory. As you move from left to right, the quantity of inventory increases; from right to left the quantity decreases.

Whether it is at the single SKU/part number or at the aggregate inventory level, there are two very important points on this curve:

▲ Point B, where we have too much inventory and there is excess cash, capacity, and space tied up in working capital.
▲ Point A, where we have too little inventory and the company experiences shortages, expedites, and missed sales.

If we know that these two points exist then we can also conclude that for each SKU/part number as well as the aggregate inventory level, there is an optimal zone somewhere between those two points. This optimal zone is depicted in Fig. 11.3.

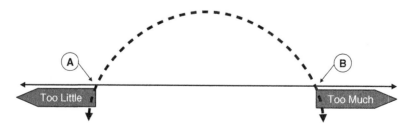

Figure 11.2   The Two Universal Points of Inventory

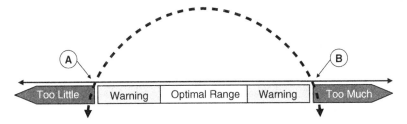

Figure 11.3   The Optimal On-Hand Inventory Range

As inventory quantity expands out of the optimal zone and moves towards point B, the return on the working capital captured in the inventory becomes less and less. The converse is also true, as inventory shrinks out of the optimal zone and approaches zero or less than zero (the typical quantity when we start to have too little). Placing point A at the quantity of zero means that inventory becomes too little when we are stocked out. Placing point A at less than zero (e.g., –1) means that inventory becomes too little when we are stocked out with demand—the definition of a true shortage.

Figure 11.4 shows the buffer spectrum with the DDMRP color-coded ranges inserted. The color-coded zones and planning algorithm are

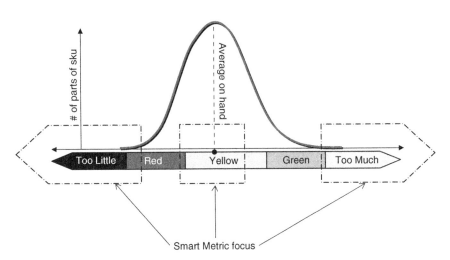

Figure 11.4   Target On-Hand Inventory and the Outliers or Tails on Both Sides of the Distribution

designed to keep the on-hand position in the optimal range. The optimal on-hand range is in the lower portion of the yellow zone. Technically, that range is defined as the top of the red zone to the summation of the top of red plus the entire green zone. The average on-hand target position is the point in the yellow zone that equates to the red zone plus half of the green zone. Shortly, we will introduce an example to illustrate this equation.

Only after the zones are understood and established can we begin to measure and manage the outliers (the tails) to drive on-hand inventory towards the middle zone. Figure 11.4 also establishes a performance measurement index to trigger buyers, planners, schedulers, operations resource managers, and deployment to take action when events drive the on-hand inventory too far outside the optimal range. All daily inventory performance is based on managing the events occurring in the tails to keep material flowing and available to meet market demand.

It is important to note that all stock buffer analysis is at the part level *not* a financial dollar value for all day-to-day measures and decisions. If a dollar value is assigned for the purposes of making a strategic investment decision, the part is valued *only* at its variable cost – its cash flow value.

Managing the impact and opportunity of the outliers is the point and the advantage of smart metrics. If market pull happens at a significantly higher rate than the current yellow zone is sized to accommodate, purchasing will be triggered to purchase more often and forced to actively manage supplier priorities against a potential stock-out position. Operations will be triggered to build more often and/or forced to continually expedite the part. If the variation is greater than the designed effective red zone protection, expedites will be triggered and/or stock-outs experienced. Both scenarios are captured in the part buffer status and direct real-time event management. Both scenarios destabilize the system and make it less reliable. These "events" also result in flow delays and reduced speed/velocity. Measuring and managing these events proactively results in achieving the four day-to-day objectives of smart metrics—system reliability, stability, speed/velocity, and focused process improvement.

Figure 11.5 shows trend reporting of parts with on-hand inventory that repeatedly enter or reside in the tails. The top graph shows parts with unacceptable service levels. It focuses on the left tail and shows the number of days over a 180-day period that each part has spent in three

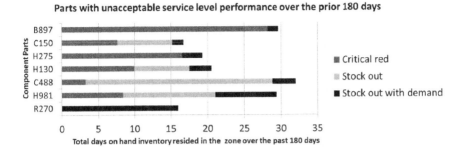

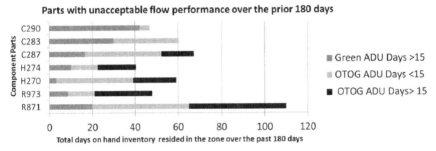

Figure 11.5 On-Hand Inventory Performances Measures Focused on Statistical Tails

different categories prioritized by the severity of the potential net loss to the system:

1. Parts stocked out with demand
2. Parts stocked out with no demand
3. Parts in the critical red (the lower half of red) zone penetration

Finance can clearly see why, where, and how much cash outflow and strategic investment is required to align the stocking levels and buffer protection to the change in demand pull and/or to protect the market from increases in supply variation. A sales review can check the product sales trend to the sales plan and signal the need for an increase in the buffer zone levels. A planning review can check if supply variability and/or time has increased and requires increased red-zone protection and/or an alternate source of supply. Regardless of the cause, these parts require an additional investment in either capacity or stock to support market targets and/or decrease expedite-related waste. Operations must act to improve the availability of these parts in order to keep the system reliable and stable and to protect the market lead-time strategy. These parts are a major source of system variation. They destabilize the system, making it less reliable, less responsive, and more wasteful. Measuring and prioritizing process improvement and strategic investment around these parts and the cause of their poor service performance achieve all of the strategic objectives of smart metrics—system reliability, stability, speed/velocity, focused process improvement, maximum strategic contribution with the minimum operating expense spend.

The bottom graph in Fig. 11.5 shows parts with unacceptable rates of flow. It focuses on the right tail and shows the number of days over a 180-day period that each part has spent in three different categories prioritized by poor flow rates:

1. Parts with on-hand inventory over the top of green with less than 15 days of average daily usage
2. Parts with on-hand inventory over the top of green that exceeds 15 days of average daily usage
3. Parts with on-hand inventory in the green zone

The choice to break categories based on an ADU greater than 15 days ensures that parts with moderate and better flow are excluded from the trend reporting. Fifteen days was chosen because it is consistent with our later examples of component part flow index categories (see Fig. 11.12 later in this chapter). Finance can clearly see the cash flow implications and working capital performance of the parts with poor flow performance:

▲ How much = the variable cost per parts × (target on-hand − actual on-hand)
▲ How long = (target on-hand − actual on-hand)/average daily usage

Over the top of green indicates the need for a sales review to check the product sales trend to the sales plan and/or a planning review of order policies and batching rules if the low-flow velocity is the result of manufacturing batches and/or purchase minimums larger than the pull rate requires. These parts are a major source of system waste. These parts reduce speed and consume cash, material, capacity, and space, and create contention for scarce resources. Reviewing buffer status trends by part and by planner provides both the benchmark to track improved performance as well as pinpoints where to focus increased investment and improvement efforts.

Setup reduction opportunities and batch size challenges for those parts using minimum order quantities (MOQ) or minimum order cycles are an integral part of the process improvement feedback loop. Parts with poor flow indexes (MOQ divided by average daily usage) are at the top of the list for review.

Measuring and prioritizing process improvement around these parts and the cause of their poor flow performance achieve all of the strategic objectives of smart metrics—system reliability, stability, speed/velocity, focused process improvement, and maximum strategic contribution with the minimum operating expense spend.

Inventory is only strategic if the market service level it provides creates a competitive advantage and/or prevents margin and share erosion. Remember, strategic inventory protects both sides of the ROI equation; maximum sales opportunity and minimum expedite-related waste focus on the left tail with a minimum invested capital focus on the right tail. Managing, measuring, and reviewing the trends of the events achieve all six of the strategic objectives of smart metrics and focus system improvement, increasing strategic contribution with the minimum invested capital and operating expense. This translates directly to increasing ROI.

The Purpose and Size of Each Strategic Stock Buffer Zone

To better understand continuous improvement strategy with stock buffers, let's use an example. Figure 11.6 demonstrates how strategic stock buffers are initially sized as a function of the average daily usage of actual past consumption, a forward-looking perspective of projected consumption, or some combination of both.

The green zone is the heart of the supply order generation process embedded in the buffer. It determines average order frequency and typical order size. It is sized either by the part's minimum order quantity (MOQ) or a percentage of the calculated average daily usage (ADU) multiplied by lead time, whichever results in a greater number. The ADU is typically calculated through the use of a rolling horizon looking backwards (e.g., the past 90 days). This percentage of ADU is determined by the lead-time category that the part falls within. The longer the part lead time, the smaller the percentage of ADU used in the equation. This is meant to force more frequent orders for long

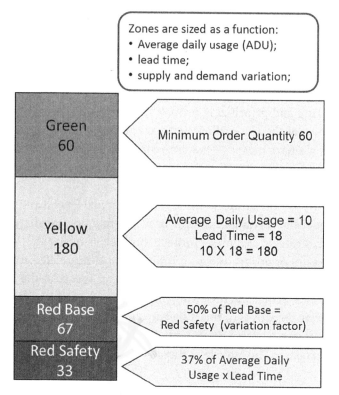

Figure 11.6 DDMRP Strategic Buffer Zones Calculated

lead-time parts, typically as frequently as order minimums will allow. This creates a conveyor belt effect as much as possible for long lead-time items. Our example part green zone is sized to its MOQ of 60 units.

The yellow zone is the heart of the coverage and shock absorption embedded in the buffer. Accordingly, the yellow zone is always set to 100 percent of ADU over lead time. Our example part yellow zone equation is: ADU of 10 × the lead time of 18 days = 180 units.

The red zone is the risk mitigation embedded in the buffer. It is a summation of two separate equations. The first equation establishes a base level (red base in Fig. 11.6) and uses a percentage of the yellow zone. Our example part red base zone equation is: 37% × 180 = 67 units.

The red safety portion of the red zone is simply a multiplier of the red zone base. This percentage establishes a safety factor and is determined by the variability category the part is placed in. The higher the variability, the higher the percentage used. In this case we are using 50 percent of the red base because the part has been placed in a midrange variability buffer profile. Our example part red safety zone equation is: 0.5×67 units = 33 units.

Figure 11.7 shows each zone quantity, purpose, and the targeted average on-hand inventory. The targeted on-hand average inventory equation is the total red zone plus one-half of the green. The example equation is: total Red (100) + (Green (60) $\times$ 0.5) = 130 units.

Buffers are dynamic because they incorporate changes in demand usage over a rolling time horizon. All inventory targets are measured in stock-keeping units, not dollar values. Any dollar value assignments should be computed with only relevant variable costs (material plus outsourcing). Variable costs provide the relevant information for short-run cost-volume-profit analysis. Variable costing of inventory generates realistic net cash flow projections tied to the required changes in strategic inventory investments based on the actual and/or anticipated change in market demand pull.

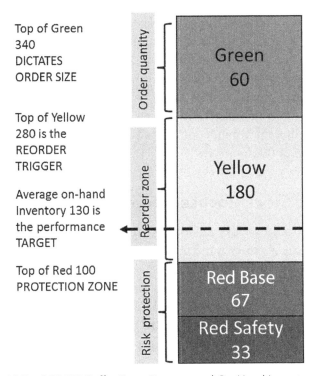

Figure 11.7 DDMRP Buffer Zone Purpose and On-Hand Inventory Target

If volumes are trending up and all other factors remain equal, then the strategic inventory investment to support the targeted market lead-time strategy must trend up as well. If not, service levels will deteriorate and expedite spending will increase. Conversely, downward trends in market demand follow the same decrease in capital investment and cash flow change. This is simply good variable budgeting, cost control, and asset performance management.

Variable cost dollar values for inventory performance targets should be computed for all strategic stock-keeping units and compared against the actual inventory on hand at the end of every month. But for day-to-day operations, buyers and planners are concerned with the stock-keeping units *not* the dollar value. Their focus is system flow. Daily reviews trigger purchase order and work order releases, and critical red buffer zone penetrations dictate expedite actions and priorities. Weekly and monthly buffer trends are reviewed at different levels of management for strategic asset assessment and action. As mentioned before, the planned average daily usage or history, or some combination, can be used to calculate the initial buffer zones. From that starting point forward, the average daily usage over lead time determines the average strategic level of inventory commitment and capacity investment commitment the company has agreed is necessary to meet its market strategy lead time and achieve its sales and marketing strategy. Purchasing, manufacturing, and distribution deployment performance are measured against the targeted market service levels and the targeted investment in strategic inventory in conjunction with the resource capacity management of the control point resources.

Strategic Stock Focused Improvements

The focus should always be–look to reduce the buffer size *without* service erosion. As seen in the previous example, strategic stock buffer zone calculations are functions of usage over time and variation. It is necessary to understand how supply lead time and variation translate directly to a change in the stock zone equations. If lead times and/or variation goes down, buffers can safely be reduced, which results in gains in flow and decreases in stored capacity, and working capital. On the other hand, if they go in the opposite direction, stock buffers must be increased *but* these increases are a necessary investment to protect the market lead-time strategy. The buffers relieve immediate pressure and give managers the time needed to focus on the causes for lead time and/or variation increases and eliminate them while maintaining market service levels.

Figures 11.8 through 11.11 use the example part from Fig. 11.7 to demonstrate how focused improvement gains translate to reductions in each of the different buffer zones and average inventory targets.

As mentioned before, our example part has a green zone set to its minimum order quantity. Figure 11.8 illustrates an MOQ reduction from 60 to 40 units and the resulting reduction in the average target inventory by 10 units. Reducing order minimums in manufacturing also releases capacity, frees up space, and speeds flow. For parts with green zones set to minimum order quantity, challenging and breaking artificial batching policies and/or supplier order minimums translates directly to a reduction in the green zone and an increase in the parts flow index. MOQs that are the result of setups to protect a scarce capacity resource will require a setup reduction effort before they can be resized to protect scarce capacity. All resources with enough protective capacity should be evaluated for setup reduction prioritized by the parts with the worst flow index (flow indexes are explained immediately following strategic stock buffer reductions).

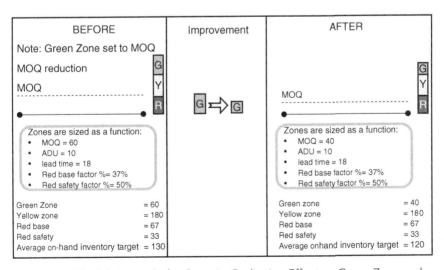

Figure 11.8 Minimum Order Quantity Reduction Effect on Green Zone and Targeted Average Inventory

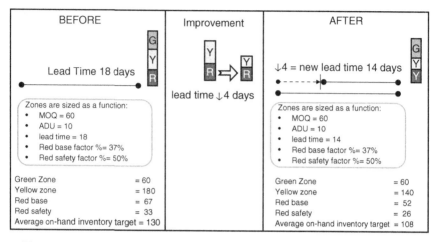

Figure 11.9 Lead-Time Reduction Effect on the Red and Yellow Zones and Targeted Average Inventory

Figure 11.9 summarizes a lead-time reduction. Insourcing, flow improvement process, or capital improvements and/or placing additional stock buffers in closer positions all can reduce lead time. This translates to a reduction in the amount of inventory protection required to ensure the targeted market service level. A lead-time reduction of 4 days for our example part translates directly to a decrease of 20 units in the yellow zone and 22 units in the total red zone. The average inventory target drops 22 units. It is important to note that this example contains a part where the green zone is set to a minimum order quantity. If the green zone was calculated as a percentage of lead time, then the green zone would also be reduced by a reduction of lead time.

Figure 11.10 summarizes a reduction in variation. The red zone safety is sized based on the volatility of both the demand and supply distortion and disruption. The red zone safety can be reduced if a more reliable source of supply is found and/or operations reduces its dependent variation events (e.g., expedites; quality disruptions; capacity contention and delays in tooling, drawings, or supplies; unscheduled maintenance; material availability) or if another stock buffer is placed closer (but still higher) in the product structure to absorb demand variability. The better and faster the events in the tails of the buffers are identified, understood, and resolved, the more stable and reliable the plan and the execution. This translates directly to the need for less red zone safety protection. In our example, the reduction in variation of 25 percent (the part variability index changed

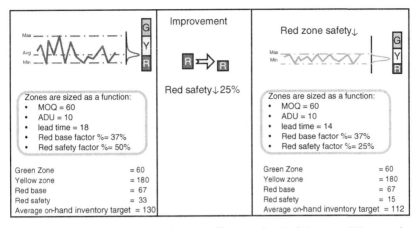

Figure 11.10 Variability Reduction Effect on the Red Zone and Targeted Average Inventory

from medium to low) translates to a drop in the red zone of 18 units. The average inventory target correspondingly drops by 18 units.

Figure 11.11 summarizes the cumulative effect of reducing all three factors over time; MOQ down 20 units, lead time down 4 days, and variability from a medium to a low designation. The result is a 32 percent reduction in the average inventory target units and the cash associated with the inventory investment. However, the gain is not limited to cash flow and working capital. Stock is stored capacity and a reduction in inventory translates to an increase in available capacity. This cycle of continuous improvement created by measuring and analyzing the events and trends in the tails of strategic stock buffers is smart metrics. Smart metrics focus people on the right improvement priorities and make the bridge directly to the effect on flow and ROI.

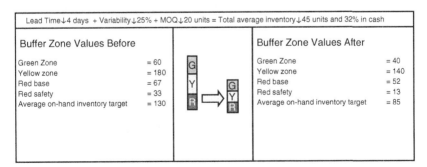

Figure 11.11 Total Reduction Effect on Buffer of Improvements in Lead Time, MOQ, and Variability

The top of the yellow zone is the reorder trigger point. In a perfect world of balanced flow and capacity, the green zone would be set to a percentage of usage over lead time based on a desired rate of flow. When significant minimum order quantities (MOQs) or minimum order cycles are present, the green zone is set to default to whichever minimum yields a larger number. This practice keeps smart metrics focused on actionable events occurring over the top of green. The assumption is the minimums exist for a valid reason such as scarce capacity. But it is important not to lose sight of the low flow performance they cause and the opportunity they represent.

The reality of most MOQs and minimum order cycles is they are not aligned to support the strategy of flow. In fact, in the authors' experience, most organizations are a target-rich environment for improvement both in reducing *and* increasing part MOQs. MOQs that are too low constantly trigger orders and can create unnecessary setups that waste scarce capacity and needlessly increase order or transactional activity. MOQs set too high block flow and waste materials, resource capacity, and cash. Calculating a part's ideal batch size and comparing it to the current batch size is a standard smart metric for assessing inventory and capacity performance and opportunity.

Flow Indexes

Figure 11.12 is a flow index summary of component parts. A flow index is created by dividing the minimum order quantity (MOQ) by the average daily usage (ADU). For example, if a part's MOQ was 60 and its ADU was 6, then the flow index would yield a score of 10, meaning that the MOQ represents 10 days of ADU coverage. Is that good? It depends on the environment and strategy. In Fig. 11.12, the first two columns define the flow index categories for Company CMG:

1 = poor, the MOQ is greater than 25 days of ADU coverage
2 = slow, the MOQ falls between 16 and 25 days of ADU coverage
3 = moderate, the MOQ falls between 9 and 15 days of ADU coverage
4 = good, the MOQ falls between 4 and 8 days of ADU coverage
5 = excellent, the MOQ is less than 4 days of ADU coverage

The higher the average demand days covered by the MOQ, the lower the flow index rating. The last column is the distribution of total parts as

Flow Index Summary - Buffered Components			
	Flow Index Range (Days)		% of Parts by Flow Index
Flow Index Rating	From	To	
Poor = 1	25	> 25 Days	11%
Slow = 2	16	25	33%
Moderate = 3	9	15	22%
Good = 4	4	8	0%
Excellent = 5	-	4	33%
			100%

Figure 11.12 Flow Index Summary—Buffered Components

a percentage of each flow index rating category. Parts with lowest flow index should be reviewed to determine if their MOQs can be challenged and flow improved.

Figure 11.13 shows the assignment of the flow index rating to each of the component parts that were used to calculate the flow index summary in Fig. 11.12. Each part's current coverage days are assigned to the matching day range in the flow index rating table in Fig. 11.12. Current coverage days are simply the number of days of ADU the current batch size will cover before triggering another work order generation.

▲ Part B897 has an ADU of 31 and a batch size of 60. On average, a work order will be released to manufacturing every 2 days. Part B897's current coverage days of 2 = a flow index rating range of 5 (excellent).

▲ Part C290 has an ADU of 2 and a batch size of 45. On average, a work order will be released to manufacturing every 90 days. Part C290's current coverage days of 27 = a flow index rating range of 1 (poor).

The ideal batch size is based on a quantity set to equal a rating of excellent flow or 4 days of current coverage days.

▲ Part B897's ideal batch size equals a quantity of 124; this is nearly double its current batch of 60 days.

▲ Part C290's ideal batch size equals 7; this is nearly 7 times less than its current batch size.

current coverage days = current batch quantity / ADU
ideal batch quantity = ADU X 4 days coverage

Flow Index for Buffered Components

Part Number	Flow Index	Current Coverage Days	Ideal Batch Quantity	Current Batch Quantity	Average Daily Usage
B897	5	2	124	60	31
C150	5	3	35	30	9
R871	2	16	10	40	3
H210	3	9	9	20	2
C290	1	27	7	45	2
C283	3	12	10	30	3
C287	2	23	9	50	2
H275	5	4	15	15	4
H270	2	19	4	20	1

Figure 11.13   The Average Demand Days Coverage Determines the Flow Index Rating

Both of these parts appear in Fig. 11.5. B897 has unacceptable service performance and C290 has unacceptable flow performance. The planner/scheduler responsible for these parts needs to evaluate the batch sizes against the capacity of the resources they consume and make the right decision based on flow. Remember, it isn't about part flow or part cost, it is about system flow and system cash generation.

Figure 11.14 is a graph of the gap between the ideal batch quantity and the current batch quantity. Again, the place to pay attention is the tails. Parts with the largest gaps should be reviewed. B897 has excellent flow but it may actually be negatively impacting the system flow. The only way to know is to understand each resource's capacity as it relates to the system capacity and system flow.

Capacity Buffers, Pareto Analysis, and Smart Metric Objectives

Figure 11.15 demonstrates a Paretian view of capacity. System capacity is best viewed as a graphical presentation where the vertical axis represents the resource capacity consumption needed to meet market demand.

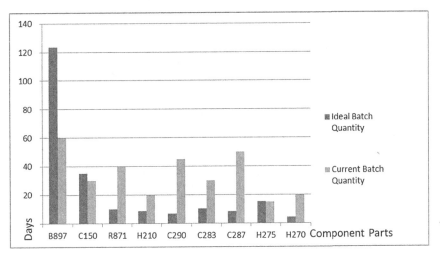

Figure 11.14 Measuring the Gap between Ideal Batch and Current Batch

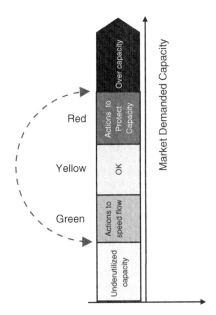

Figure 11.15 Color-Coded Zones Indicate the Resource or System's Ability to Respond to Market Opportunities Inside Market Lead Time

The color-coded zones represent the resource or the system's ability to respond to market opportunities inside lead time.

Resources with capacity demand consistently in the bottom two (blue and green) zones have excess capacity and represent opportunities to speed flow. Resource management actions should be focused on flexibility and agility, batch size reductions, cross training, and so on. In the reality of complex adaptive systems, these resources define an organization's resiliency, the ability to recover quickly and stabilize flow. These resources also define the opportunity to find markets and/or products for strategic contribution that require no additional investment. *Do not artificially increase utilization of these resources beyond true demand pull to make a "dumb metric" look better.*

Resources with capacity demand consistently in the middle (yellow) zone have capacity to respond to the system's current demand-and-supply variation inside market lead time.

Resources with capacity demand consistently in the red zone have limited ability to respond to variation. They may need to be protected from variation with some combination of strategic stock and time buffers. Resources in the upper half of this zone are candidates for strategic control points.

Resources with capacity demand consistently in the top zone are the top priority for process improvement (setup reduction, cross training, scrap reduction, etc.), outsourcing, or capital acquisition. They *are* strategic control points and need to be protected with both strategic stock and time buffers. These resources determine the flow rate of the entire system. The rate they generate strategic contribution is the rate the system generates cash flow and profit.

Our Reference Environment

It is time to introduce the company case we will be using to build understanding and create specific and meaningful examples for the rest of the chapter. Before analyzing resource capacity, we introduce our reference Company CMG and their demand driven system design. Figure 11.16 is the demand driven design previously discussed in Chap. 3 with one modification: we have chosen to include heat treat inside component manufacturing.

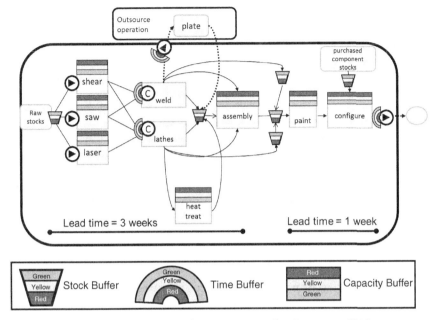

Figure 11.16 Demand Driven Design for Company CMG

▲ Stock buffers are strategically placed to compress lead times and decouple variation from being passed from one process area to another process area and into the market. The longest lead time chain of unbuffered dependencies is three weeks for resupplying component parts to assembly and one week for resupplying finished items through assembly.

▲ Control points have been chosen at weld and lathes and they are both protected with time buffers. Outsourcing to plate and shipping are both protected with time buffers to protect their delivery schedules.

▲ All other resources have sufficient capacity over and above weld and lathe to respond to variation and reliably deliver on time to the scheduled capacity rate of weld and lathe.

The System and Paretian View of Resource Capacity

The starting point to managing resources is to understand their capacity in relationship to meeting the market demand strategy. Stock can be converted to capacity, and the sales plan compared against capacity.

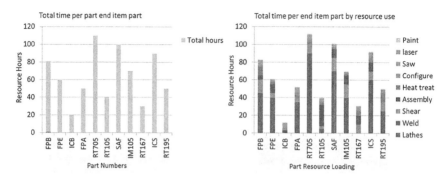

Figure 11.17 Converting a Part Quantity to Resource Capacity

Each part routing has a resource time that can be translated to a total process time and to individual resource required capacity process time. Figure 11.17 demonstrates the different resource requirements for 11 individual end item parts. Clearly different parts have different resource load requirements as well as differences in the total resource time they require.

Total time per part tells us little. It requires a system Paretian view to understand the strategic resource commitment and the volume restrictions inside Company CMG's current resource capital and manpower investment limits. Figure 11.18 puts strategic capacity management and investment in perspective. Graphs of aggregated product demand and the

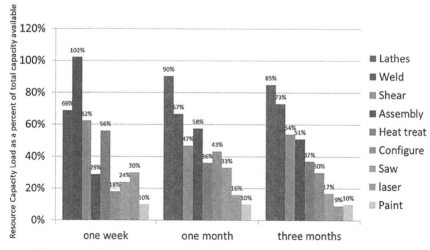

Figure 11.18 Resource Loading over Three Different Time Horizons

capacity used by all resource areas over different time periods are displayed. The graphs demonstrate an emerging pattern over time, as well as short run (one week) resource contention.

The tails of the distribution provide a clear view to the resources with sufficient protective capacity as well as the resource with potential limited capacity to output. Again, it is all in the tails. Comparison of the three resource capacity utilization graphs clearly shows the short-term demand variation and the smoothing effect of aggregation over time. Clearly lathes has the highest resource demand against available capacity over time and welding has short-term demand spikes that can create short-term capacity contention that must be managed. Every other resource has enough protective capacity.

The Paretian view of resource capacity confirms lathes and weld are strategic control points. Lathing and weld, respectively, meet all of the criteria for good subsystem control points:

1. Together they control the flow of all parts prior to assembly.
2. They are both points of scarce capacity relative to all other resources.
3. They both require scarce skilled labor sets.

Strategic control points define relevant information for both strategic investment and day-to-day execution decisions. These control points have been strategically chosen to break variation, control and focus feeding resource process priorities, and maintain schedule coherence.

Weld and Lathe capacities effectively determine the total potential volume output. They are scheduled to demand pull based on their available finite capacity prioritized to the intermediate stock buffer zone statuses they feed. Finite capacity scheduling of control points is the only way to ensure a reliable and executable schedule. When there is more demand than components can deliver, Sales must choose the customers or products with the highest priority status until more capacity can be acquired. The supply chain must have a plan to handle both temporary surges in capacity as well as when to invest in permanent increases in fixed capacity. Strategic buffers between components and assembly must be sized to absorb the temporary surge in demand shown in the one-week graph of resource capacity required in Fig. 11.18. Remember, if the strategic stock buffers collapse both Manufacturing and Distribution will be hit with disruption. As a result, variation, expediting, and the associated costs increase; service levels fall; and ROI declines.

The Implications of the Sales and Operating Plan for Strategic Capacity Investment

Strategic stock buffers are designed to be dynamic and adjust zone sizes based on actual ADU changes over a chosen time horizon. Normal supply-and-demand variation is covered in the red zone sizing. Planned events or known seasonal fluctuations require a planned increase or decrease in capacity to build strategic stock buffers a time frame ahead of the change in demand pull. Strategic stock buffers are designed to protect schedule reliability and market performance and can be adjusted up or down ahead of and after changes in product demand. Control points that do not have enough protective capacity to build ahead must be either offloaded with strategic outsourcing or there must be a capacity investment that will increase both fixed cost and output volume range so they can reliably deliver the increase to the strategic stock buffers on time.

Figure 11.19 shows the successive wave backwards in time through the supply chain of the rolling 13-month sales plan. For each known or planned event, the end item products and the strategic stock buffers they draw from (assembled, machined, purchase parts) must each be adjusted a time frame ahead of each other. There must be enough component capacity (weld and lathe) to supply the increased demand or the stock buffers they feed must be built ahead of time at the rate their protective capacities can reliably deliver.

Operations must answer these questions with regard to weld and lathe capacity:

▲   Is there enough capacity to respond to both short-term variations in demand and product mix?
▲   If not, how much capacity is needed, when, and what resource?
▲   What is the plan to provide sufficient capacity to meet the sales plan inside the market lead time strategy?

Figure 11.20 shows weld can be managed through the required production volume increase with a combination of building strategic stock ahead of the demand pull and with planned overtime. However, lathe has no surge capacity to build ahead. Capacity will need to be secured prior to March to ensure the strategic stock buffers are sufficient to meet the increase in market demand. If lathe capacity is not increased, Sales must decide how to allocate scarce capacity between customers and products. Remember, in a nonlinear reality of connected interdependencies, it is the

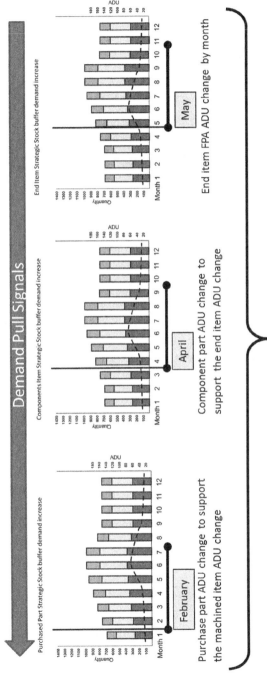

Figure 11.19 Sales Plan Volume Increase in May Requires Strategic Buffers to Increase Ahead of Market Pull

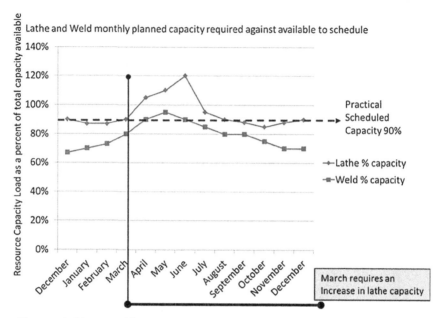

Figure 11.20 Monthly Capacity Increases at Weld to Meet Projected Monthly Sales Demand

resources in the tails that determine strategic contribution because they limit the supply output.

Cost, Volume, Profit Relevant Range Is Defined by the Scarce Capacity Resource

Figure 11.21 demonstrates how the scarce capacity of lathe is the volume-limiting factor of the system in the cost-volume-profit graph. Lathe capacity defines the volume-relevant range as well as the fixed-cost range. It is easy to quantify the current potential of the supply chain given the current limitations of the existing asset base and product mix and product contribution.

Allocating Scarce Capacity to the Market Based on Strategic Contribution

The first step in calculating strategic cash flow is to understand how the rate of product flow relates to the rate of cash flow. Total volume output is governed by the scarce capacity resource, lathes. All other resources have

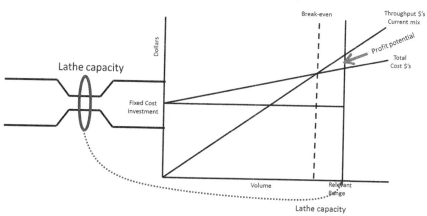

Figure 11.21   Lathe Capacity to Define the Cost-Volume-Profit and Fixed-Cost Relevant Range

enough capacity to meet the expected sales plan by period and enough surge capacity to rebuild their stock buffers and protect demand flow from supply-and-demand variation. Figure 11.21 shows the rate of strategic contribution commonly known as the throughput dollar rate. The relevant information to determine product profitability must consider the cash generation per unit of the scarce resource. Decisions on how to ration capacity to products and customers require understanding the tradeoffs in net cash flow and net profit between the different products that use lathe capacity. This is a well-known concept in management accounting and is commonly described as contribution margin per unit of the constrained resource:

> If some products must be cut back because of a constraint, the key to maximizing the total contribution margin may seem obvious—favor products with the highest contribution margins. Unfortunately, that is not quite correct. Rather, the correct solution is to favor the products that provide the highest contribution margin per unit of the constrained resource.[1]

Figure 11.22 clearly shows how product FPA generates cash more than 8 times faster than product RT195 and 5.4 times faster than Product SAF

---

[1]Noreen, Eric, Peter Brewer, and Ray Garrison, "Managerial Accounting for Managers," McGraw-Hill, 2008, page 516.

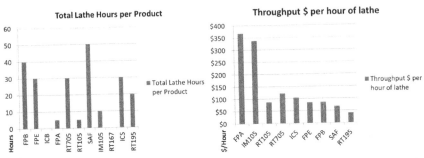

End item product throughput dollar rates are calculated based on the rate they generate cash per lathe hour										
End Item Part	FPA	IM105	RT105	RT705	ICS	FPE	FPB	SAF	RT195	
Selling Price	$2,550	$4,500	$540	$5,400	$3,900	$3,600	$4,500	$4,800	$1,050	
Total Variable Costs	$714	$1,125	$108	$1,728	$780	$1,080	$1,125	$1,440	$263	
Throughput/Contribution $	$1,836	$3,375	$432	$3,672	$3,120	$2,520	$3,375	$3,360	$788	
Lathe hours	5	10	5	30	30	30	40	50	20	
Throughput $ per hour of lathe	$367.20	$337.50	$86.40	$122.40	$104.00	$84.00	$84.38	$67.20	$39.38	

Figure 11.22 Calculating Strategic Contribution = Throughput Dollar Rate of the Lathe Scarce Resource

even though SAF's dollar contribution margin is nearly twice as large as FPA's. Remember, the definition of relevant information is the cost and benefits that differ between alternatives. From a financial perspective it is the net cash flow difference between the alternatives. In order to protect strategic customers, Sales may have to make a different choice but the lost opportunity should be quantified, known, and understood. They might even change their definition of who are the strategic customers.

Figure 11.23 shows the difference in total net profit and cash flow if the products with the best throughput dollar rate per unit of lathe time are given preference over the products with the worst throughput dollar rate. If more lathe capacity is not found for the four months when demand exceeds capacity, profit and cash flow can only be maximized by prioritizing the allocation of machined parts to products with the highest throughput rate. The assembly build schedule would be prioritized by the finished product in the order of precedent in Fig. 11.22. The assignment of scarce machined parts by prioritizing sales orders for FPA then IM105, RT105, RT705, ICS, FPE, FPB, SAF, and lastly RT195 will ensure the highest rate of strategic contribution for the entire invested capital of the system. Of course, Sales needs to understand the allocation and make their decision based on

The lost opportunity of allocating the capacity shortfall to products with the lowest rate of dollar throughput

End Item Part	Throughput $ per Hour of Lathe	April	Monthly Hours	May	Monthly Hours	June	Monthly Hours	July	Monthly Hours	Total Dollars	Total Hours
FPA	$367.20	$36,720	100	$36,720	100	$36,720	100	$36,720	100	$ 146,880	400
IM105	$337.50	$47,250	140	$47,250	140	$47,250	140	$31,050	92	$ 172,800	512
RT705	$122.40	$14,076	115	$14,076	115	$14,076	115	$0	0	$ 42,228	345
ICS	$104.00	$22,984	221	$40,560	390	$40,560	390	$0	0	$ 104,104	1,001
RT105	$86.40	$0	0	$1,987	23	$35,165	407	$0	0	$ 37,152	430
FPB	$84.38	$0	0	$0	0	$0	0	$0	0	$ -	0
FPE	$84.00	$0	0	$0	0	$0	0	$0	0	$ -	0
SAF	$67.20	$0	0	$0	0	$0	0	$0	0	$ -	0
RT195	$39.38	$0	0	$0	0	$0	0	$0	0	$0	0
Totals		$121,030	576	$140,593	768	$173,771	1,152	$67,770	192	$ 503,164	2,688

The lost opportunity of allocating the capacity shortfall to products with the highest rate of dollar throughput

End Item Part	Throughput $ per Hour of Lathe	April	Monthly Hours	May	Monthly Hours	June	Monthly Hours	July	Monthly Hours	Total Dollars	Total Hours
RT195	$39.38	$17,325	440	$17,325	440	$17,325	440	$7,560	192	$ 59,535	1,512
SAF	$67.20	$9,139	136	$22,042	328	$26,880	400	$0	0	$ 58,061	864
FPE	$84.00	$0	0	$0	0	$26,208	312	$0	0	$ 26,208	312
FPB	$84.38	$0	0	$0	0	$0	0	$0	0	$ -	0
RT105	$86.40	$0	0	$0	0	$0	0	$0	0	$ -	0
ICS	$104.00	$0	0	$0	0	$0	0	$0	0	$ -	0
RT705	$122.40	$0	0	$0	0	$0	0	$0	0	$ -	0
IM105	$337.50	$0	0	$0	0	$0	0	$0	0	$ -	0
FPA	$367.20	$0	0	$0	0	$0	0	$0	0	$0	0
Totals		$26,464	576	$39,367	768	$70,413	1,152	$7,560	192	$ 143,804	2,688

$359,3600 is the total difference in net profit and net cash flow between the optimum product mix and the worst product mix

Figure 11.23 The Right Product Mix Choice is Worth $359,360 in Net Income and Net Cash Flow

the product mix per strategic customer. All metrics and incentives must be aligned to maximizing strategic contribution of the system to avoid conflicting decisions and actions that jeopardize system coherence.

Prioritizing the Product Mix by Throughput Dollar Rate Changes the CVP Graph

Figure 11.24 uses CVP to graphically demonstrate the positive effect on both breakeven and total net profit potential when scarce capacity is allocated first to the products with the highest throughput dollar rate per unit of lathe time. Using the targeted product mix in the previous figure changes the slope of the total revenue line and shifts the break-even volume to the left, generating a maximum increase of $359,360 in strategic contribution. There is simply a greater cash flow per unit of constraint time generated with the new mix. Throughput dollar rate based exploitation is a mix manipulation strategy that is only effective when there is a scarce capacity resource that either cannot be broken, or the company chooses not to increase fixed investment to break the bottleneck. If the cost of increased capacity exceeds the short-term value of the increase in profit potential, companies are better off finding either temporary outsourcing or the best profit maximizing strategy in the short run.

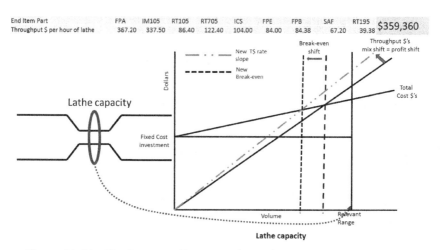

Figure 11.24 The Positive Effect on Both Break-Even and Total Net Profit from Shifting to a Product Mix driven by Throughput Dollar Rates Per Unit of Lathe Time

Calculating the Value of Finding More Lathe Capacity

Regardless of how scarce capacity is allocated, strategic contribution per product or strategic contribution per customer product mix, Manufacturing will not be able to fulfill the total planned demand at its current level of lathe capacity. It is Operations' job to know the relevant information, Finance's job to know the right numbers, and Sales' job to make the right strategic market choice. Smart metrics are focused on the few control and decoupling points that highlight the relevant information and right numbers for everyone in the system. Figure 11.25 shows the strategic contribution of the products and the volume shortfall in the sales plan due to the lack of lathe capacity. The $263,283 decrease will decrease both net profit and cash flow. Operations should immediately evaluate both outsourcing and increasing lathe capacity through a capital acquisition. Note in this example we are using the contribution dollars for each end item product because the assumption is we can and will make a strategic investment in lathe capacity and capture all available market opportunity.

End Item Part	FPA	IM105	RT105	RT705	ICS	FPE	FPB	SAF	RT195	
Selling Price Unit	$2,550	$4,500	$540	$5,400	$3,900	$3,600	$4,500	$4,800	$1,050	$30,840
Total Variable Costs Unit	$714	$1,125	$108	$1,728	$780	$1,080	$1,125	$1,440	$263	$8,363
Throughput/Contribution Unit	$1,836	$3,375	$432	$3,672	$3,120	$2,520	$3,375	$3,360	$788	
Total product increase	9	9	24	13	11	14	11	13	10	
Total Contribution Increase	$16,524	$30,375	$10,368	$47,736	$34,320	$35,280	$37,125	$43,680	$7,875	$263,283

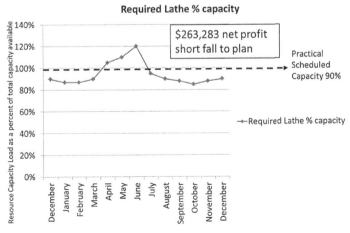

Figure 11.25 Strategic Contribution Shortfall Due to Lathe Capacity Shortfall

Throughput dollar rates are only applicable when we need to make a choice on how to allocate scarce resource capacity, material, space, and so on.

Relevant Information for Strategic Outsourcing

An outsource decision should always be based on the need to free up scarce capacity to meet market demand. The only exception would be a part with a variable cost that is greater than the purchase part cost. To date, the authors have not experienced that reality. The only other caveat is the assumption that the existing internal process produces a part that meets both market quality specifications and expectations. Outsourcing decisions have a financial impact from both the change in variable costs and the change in working capital investment in the strategic stock buffers. Any other known changes in operating expense such as space or scrap rates must be accounted for as well. Choosing which parts to offload and selecting the outsource vendors require a mix of financial and nonfinancial relevant information. Determining which parts are best to outsource requires understanding the difference in the relevant variable cost between make or buy and the investment change required in the strategic stock buffers of the outsourced part candidates due to increased lead time.

The starting point is to identify common component parts of the end items with the largest lathe capacity requirements. These are the parts that will free the most hours of lathe capacity and most likely result in the minimum investment in aggregated strategic stock. Figure 11.26 is a diagram of the end item parts with the largest lathe capacity requirements, the identified common stock buffer position, and their lathe part components. End item SAFs have the largest requirement for lathe capacity and component part C290 requires three separate lathe parts.

Figure 11.27 shows the total variable cost of outsourcing the capacity shortfall of lathing is $161,000 with a positive strategic contribution of $102,183. There will also be a temporary increase in the stock buffers for the outsourced parts B897 and C290 to cover the increase in lead time required to move the parts to and from the outsource operation. The ADU for the end item parts FPE, ICS, FPB, and SAF must be covered in each of the stock buffer zones over the three days of additional lead time required to outsource. All relevant information to understand the financial cost and benefits of outsourcing is detailed in Fig. 11.27. Relevant information is always limited to what differs between alternatives.

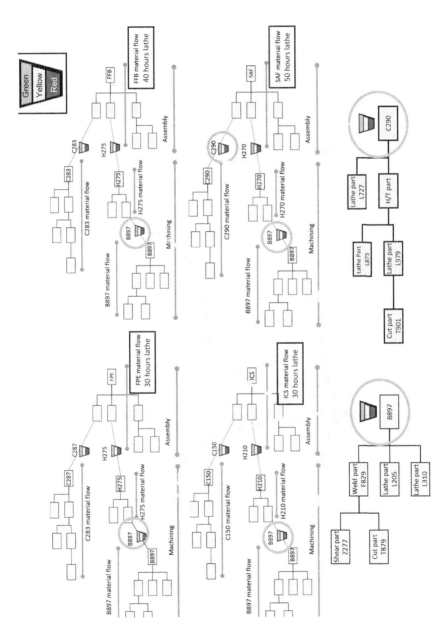

Figure 11.26 Component Parts with the Largest Aggregate Lathe Capacity Utilization

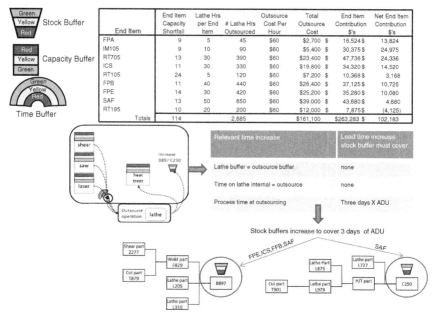

End Item	End Item Capacity Shortfall	Lathe Hrs per End Item	# Lathe Hrs Outsourced	Outsource Cost Per Hour	Total Outsource Cost	End Item Contribution $'s	Net End Item Contribution $'s
FPA	9	5	45	$60	$2,700	$ 16,524	13,824
IM105	9	10	90	$60	$5,400	$ 30,375	24,975
RT705	13	30	390	$60	$23,400	$ 47,736	24,336
ICS	11	30	330	$60	$19,800	$ 34,320	14,520
RT105	24	5	120	$60	$7,200	$ 10,368	3,168
FPB	11	40	440	$60	$26,400	$ 37,125	10,725
FPE	14	30	420	$60	$25,200	$ 35,280	10,080
SAF	13	50	650	$60	$39,000	$ 43,680	4,680
RT195	10	20	200	$60	$12,000	$ 7,875	(4,125)
Totals	114		2,685		$161,100	$263,283	102,183

Figure 11.27 The Impacts of the Outsource Plan (Variable Cost Increase, Positive Net Contribution, Lead Time Increase, and Increased Inventory Investment)

The Case for Acquiring Internal Lathe Capacity

Resource capacity at lathing is constrained but lathing is not one homogenous pool of capacity. Figure 11.28 is a drill-down of the different lathe resources and their different capabilities that make up the total lathe capacity. There are currently 12 lathes. Five are lathes capable of handling small parts. Four are capable of handling small and medium parts, and three are capable of handling medium and large parts. Both the small and large lathes have some protective capacity (10 percent plus overtime) to recover from short-term variation. Medium lathe capacity is currently fully scheduled and is using overtime to cover its normal schedule. It is dependent on the large lathes' ability to offload work as its only form of protective capacity. Any additional demand for medium lathes will seriously erode the large lathes' protective capacity. Obviously, if we are going to outsource parts it would be best to concentrate on medium lathe parts. Just offloading lathe in general would add cost without solving the

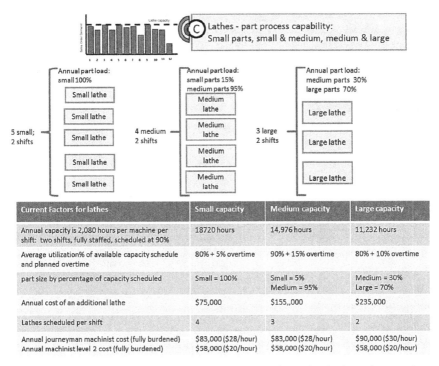

Figure 11.28 Annual Lathe Resource Capacity by Individual Machine and Process Capability

capacity problem if we fail to understand the source and reason for the capacity limitations in lathes. There is another alternative to review that increases lathe capacity; invest in one more medium lathe.

Figure 11.29 is a summary of the increase in operating expense and capital investment necessary to capture the $263,283 in contribution (throughput dollars) at risk due to insufficient lathe capacity. After covering the relevant increase in operating expense, the net increase in contribution is $156,483 and results in a permanent increase in medium lathe capacity to cover the rest of the sales plan year. The $155,000 cost of the capital for the additional lathe is fully recovered by the additional contribution in the first eight months of the plan year.

Figure 11.29 also shows a graph of the increase in lathe capacity compared to the previous available capacity. With the additional lathe, there may be just enough protective capacity to allow strategic stock to be built ahead of the increase in demand pull. Acquiring medium lathe capacity is a

Relevant cost and benefits for lathe capital acquisition	cash inflow and (cash outflow)
Capacity increase (2 shifts @ 90%) covers the lathe shortfall of 2,700 hours	3,744 hours
Decrease ins overtime 1,200 hours @$36/hour	$43,200 Operating expense↓
Add two machinists level 2 (one per shift)	($150,000) Operating expense↑
Total strategic contribution of meeting planned sales targets	$263,283
Net operating cash flow change for the plan year	$156,483
Cost of capital acquisition is recovered inside the plan year	$155,000

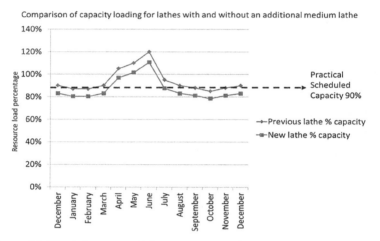

Figure 11.29 A Summary of the Increase in Relevant Operating Expense and Capital Investment to Cover the Four-Month Capacity Shortfall

necessary investment to cover the shortfall but more importantly to regain surge capacity and flexibility at the large lathe. The medium lathe is already using permanent overtime to cover its normal load and 30 percent of the large lathe capacity, a much larger capital investment. No other financial information in our supply chain example is relevant to make the decision between outsourcing and acquiring internal lathe capacity.

A Summary of the Financial and Nonfinancial Factors of Outsourcing or Capital Investment

The real question is the longer term market outlook. If there is an expectation that sales will continue at its current rate, a permanent solution for medium lathe capacity must be found. A side-by-side comparison of the two alternatives is provided in Fig. 11.30.

Factors	Outsourcing	Lathe investment	Net annual difference
Annual increase lathe hours	2,685	3,744	1,059
Annual variable cost increase	$161,000 ($60 rate)	$(0)	$161,000
Annual variable cost increase for equal hours	$63,540	$(0)	$63,540
Net Increase in labor cost	$0	$(106,800)	$(106,800)
Net annual cost increase of outsourcing	$224,640	$(106,800)	$117,840
Monthly net cost increase of outsourcing			$9,820

Capital Investment Factors

Cost of capital	$155,000
Monthly net cost increase	$ 9,820
Payback on cost of capital	< 16 months
One time permanent investment in working capital for inventory increase ≈ $12,000	

Figure 11.30 Comparison of Outsourcing and Lathe Investment in Capital and Labor

The annual variable cost increase of $161,000 only secures 2,685 hours of capacity. Purchasing a medium lathe secures 3,744 hours of capacity with an annual operating expense increase of $106,800 (the net labor increase). The cost of 3,744 hours in outsource capacity at $60 per hour is $224,640. The net cash flow difference between outsourcing and a capital investment is $117,840, or $9,800 a month. The cost of the lathe has a less than 16-month payback. Our comparison ignored the necessary increase in inventory required to support the outsourcing alternative. Our purpose is to provide examples of relevant information and we are ignoring the net tax effects of depreciation and interest. Finance can add any applicable cost of capital factors—what they should not do is add a GAAP spin. The above translates directly to flow, product, and cash, and the required investment in real dollars to secure and sustain flow at the rate of market demand and the organization's market lead-time strategy.

Figure 11.31 is a summary of equally important nonfinancial factors to consider when making a strategic decision. Organizations have very different scarce resource issues and they can shift over time. The authors often find space a limiting factor in outsourcing decisions if the increase in stock is large or the parts outsourced are used in a large strategic component part. A list of both the limiting system factors and the strategic concerns is useful to capture both the type of issue and its relative importance to system flow for strategic outsourcing decisions. Protecting the market lead-time strategy will require a higher on-hand quantity to absorb the difference in both lead time and supplier variability. There are usually hidden costs associated with the least cost supplier. If the part is small, feeds a low-variable-cost strategic stocking part, and has low risk of quality

Part number	Technology risk	Quality risk	Space	Variable cost of strategic stock	MOQ/ADU flow index	Market lead time	Supplier Reliability
L205	1	1	1	1 (B897)	5		
L310	1	1	1	1 (B897)	5		
L875	1	1	1	3 (SAF)	2		
L979	1	2	3	3 (SAF)	2		
L727	3	3	1	3 (SAF)	2		

Risk factor inherent in the part type:
1= low
2= medium
3= high

Supplier dependent and must be considered in conjunction with the cost and MOQ of the different sourcing options

Nonfinancial factors are for the child parts of these components

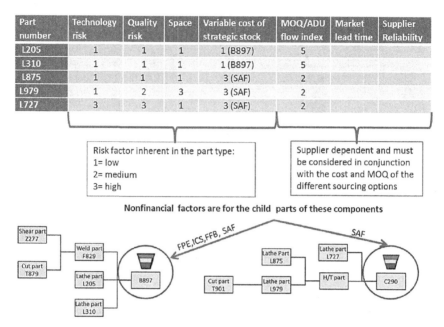

Figure 11.31 Nonfinancial Factors to Consider in Outsourcing Decisions

or technology loss, then it is a good candidate to source based on cost. The higher level of investment in working capital to protect against a supplier with a longer lead time but better cost can be an acceptable trade-off as long as the MOQ still provides an acceptable flow index.

The Third Alternative—Using Smart Metrics to Find Capacity

Earlier in this chapter we introduced the ideal batch size based on the assignment of a flow rate index of four days of coverage for any component part's minimum order quantity. Setting batch sizes to no greater than a four-day rate of ADU coverage would average a work order release every four days. Previously, Fig. 11.14 showed a graph of each component item's ideal batch quantity to its current batch quantity and the gap comparison. Three component items had positive gaps where the current batch quantity was less than the ideal: B897, C150, and H275. Additionally they have been trending with unacceptable service levels. Figure 11.32 shows all three parts have experienced stock-outs with demand and critical red over the past 180 days.

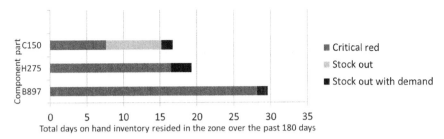

Figure 11.32 Service-Level Performance of C150, B897, and H275 over the Previous 180 Days

Component parts experiencing stock-outs and critical red zone penetration trigger expedites in component operations. Parts experiencing stock-outs with demand delay the schedule of assembly end items and negatively impact reliable market service in addition to the expedite cost they incur. These parts' service levels will benefit and we can release capacity stored in excess setups at our scarce resource by increasing their MOQs.

Figure 11.33 illustrates each of these component parts, their respective child parts that pass through lathes, and the end item products they each supply. The table embedded in Fig. 11.33 shows the total number of lathe

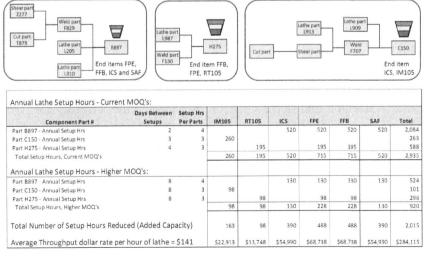

Annual Lathe Setup Hours - Current MOQ's:									
Component Part #	**Days Between Setups**	**Setup Hrs Per Parts**	**IM105**	**RT105**	**ICS**	**FPE**	**FFB**	**SAF**	**Total**
Part B897 - Annual Setup Hrs	2	4			520	520	520	520	2,084
Part C150 - Annual Setup Hrs	3	3	260						263
Part H275 - Annual Setup Hrs	4	3		195		195	195		588
Total Setup Hours, Current MOQ's			260	195	520	715	715	520	2,935
Annual Lathe Setup Hours - Higher MOQ's:									
Part B897 - Annual Setup Hrs	8	4			130	130	130	130	524
Part C150 - Annual Setup Hrs	8	3	98						101
Part H275 - Annual Setup Hrs	8	3		98		98	98		296
Total Setup Hours, Higher MOQ's			98	98	130	228	228	130	920
Total Number of Setup Hours Reduced (Added Capacity)			163	98	390	488	488	390	2,015
Average Throughput dollar rate per hour of lathe = $141			$22,913	$13,748	$54,990	$68,738	$68,738	$54,990	$284,115

Figure 11.33 Increased Lathe Capacity Available from Raising the Component Parts Batch Size Minimums to 8 Days of Coverage

hours released to run parts by increasing the minimum batch sizes and reducing the number of setup hours accordingly. Increasing the current batch size to 8 days of coverage drops the flow index from excellent to good (5- to 8-day coverage range) for each of the component end item parts but will release 2,015 hours of capacity previously consumed by the extra setups. Note, the objective has nothing to do with a lower unit cost and everything to do with flow through the scarce resource and the flow of the system it governs. The average throughput dollar rate per unit of lathe, at the current product mix is $141 that equates to $284,115 dollars of potential cash and net profit increase without any additional increase in capacity or labor. Just as important it creates system stability and reliability for a minimal working capital investment.

To many lean practitioners, raising batch size minimums would appear to be heresy and counter-intuitive to flow. But in the world of nonlinear systems, it is simply good logic. The flow index exists as an information point, not a rule or end point to drive to or a competing metric to achieve. The truth is that by lowering flow on certain items we get more total system flow. Smart metrics are smart because the information they provide is always interpreted in light of the current state of the stock buffers and the control points provided by the feedback system. In fact, it was the combination of evaluating poor component part service performance trends in Fig. 11.32 and the scarce resource capacity availability that created the system improvement process focus.

Strategic stocked part batch minimums must consider the resources with limited capacity and a balance between system flow and individual component part flow. The company sells and ships end items, not component parts, and must have all of the child component parts available in order to assemble and ship a single finished item. If the company does sell component replacement parts, their availability is even more critical. For these reasons the authors do not recommend increasing the size of the batch beyond the good flow index strategically determined. Any additional lathe capacity required must be attained without negatively impacting the system's flow and strategic market lead time.

Figure 11.34 is a CVP graph explaining the net profit potential available due to decreasing component part minimums and increasing scarce capacity without adding to fixed cost (additional lathe and labor) or variable costs (outsourcing). The relevant range is defined by the total capacity of lathes.

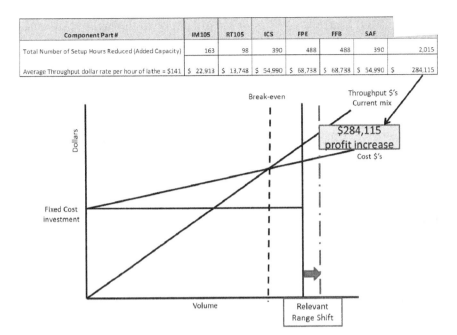

Component Part #	IM105	RT105	ICS	FPE	FFB	SAF	
Total Number of Setup Hours Reduced (Added Capacity)	163	98	390	488	488	390	2,015
Average Throughput dollar rate per hour of lathe = $141	$ 22,913	$ 13,748	$ 54,990	$ 68,738	$ 68,738	$ 54,990	$ 284,115

Figure 11.34 Expanding the Relevant Range by Saving Setups

Expand lathe capacity and the relevant range shifts. The potential profit increase is stated as the average dollar throughput rate per lathe hour but if Sales can focus the increase on the right product mix (Fig. 11.22), the net profit increase could double. Sales incentives should always be tied to dollar throughput rates of products at the scarce resources and control points with the least capacity where additional investment is high and or risky. Sales incentives should not be based on total volume, gross sales, or gross profit GAAP values—they have little to no relationship to cash and profit velocity or ROI. The range of throughput dollars per unit of lathe hour is $39/hour to $367/hour for the same resource capacity investment; a multiple of 9 times in cash and profit velocity. In our Company CMG example the range in throughput dollar rates translates to a potential cash and profit increase from $1.8 million to $16.5 million dependent on the product mix in relation to lathe capacity usage.

Sales planning, marketing, and incentive systems need to be aligned to products that generate the highest profit velocity. A shift to either side of the spectrum means the difference between a dramatic net loss or a dramatic profit and ROI increase. A 9× multiple difference in throughput dollars/time is not unusual. The authors have experienced product throughput dollar rates with multiples as low as 3× ($6,000/hour to $18,000/hour at a steel mill) to multiples of 19× ($80/minute to $4/minute at a circuit board manufacturer). Although the multiple was smaller at the steel mill, the sheer size of the dollar spread resulted in capturing huge gains in profit velocity. Both companies used this smart metric to align their market strategy and dramatically leverage their resource investments with significant gains in profit velocity, cash flow, and ROI.

Summary on Strategic Investments in Capacity and Stock

Capacity and stock are both investments in time. They are interdependent and to a large extent interchangeable. Investments in stock or capacity are only strategic if they protect and deliver the market lead-time strategy. The product strategy needs to align or at a minimum understand and consider the current resource capacity relationship to cash and profit velocity to each of these products. It is easy and common to find plants filled to capacity with a high-margin but low-profit velocity product mix. In the author's experience, these two are never the same product mix and if it were to occur, it would be a temporary accident, not a managed event. The exception to this rule is the world of commodity products. True commodity products are rare in today's world but they do still exist. We only have to look at the example of Henry Ford to clearly see that when the automobile was a commodity product, Ford had a lock on the market with a one-product, high-flow, least-cost strategy. Sloan and F. Donaldson Brown recognized the change in the consumer market and literally drove Ford from the market with a product niche strategy made possible by understanding CVP analysis. Brown's methods and CVP formulas are still relevant today as long as the relevant information used in the analysis is defined by nonlinear system rules and Paretian math.

The Scheduling Implications of Stock Buffers

There are two scheduling implications when strategic stock buffers are used to decouple different plant processes. The first is demonstrated in

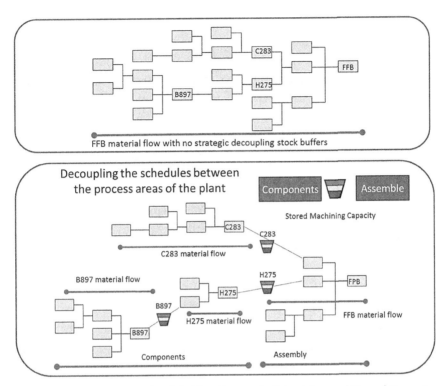

Figure 11.35   Strategic Stock Buffers Decouple Schedules and Stored Capacity

Fig. 11.35. Strategic stock buffers are used to create shorter independent planning and execution horizons. Breaking dependencies between processes means not only decoupling lead times but decoupling the scheduling equation and dependencies of one area from another. This translates to an easier and more stable scheduling scenario. The result is an increase in due date performance reliability. Stock buffers store capacity and add protection to scarce capacity resources. The common stored capacities of resources contained in stock buffers provide aggregated protection against fluctuations in demand. They act like a shock absorber and take the pressure and urgency out of short-term variation events. This stored protective capacity provides planners and schedulers with more flexibility when there is short-term capacity or material contention. Well-managed strategic stock buffers create stable, reliable planning horizons and more flexibility for schedulers to protect both scarce resource capacity and due date performance to market expectation.

The second implication for scheduling is the use of stock buffer status to provide a demand pull scheduling signal. Stock buffers' visible zone status provides clear priorities for purchasing, planning, execution, and supply deployment. Visibility to real-time strategic stock status is a necessary condition for a relevant feedback loop and smart metrics. Coherence to real market pull generates work order priorities from the simple, visible signals using zone color and buffer percentage remaining. Trending the tails of the buffers provides performance assessment on the use of cash, capacity, space, inventory, and market service level performance. Examples of the importance of using buffer zone trend analysis were discussed in the above sections. Figure 11.36 demonstrates the use of stock buffer zone status to determine both the work order quantity generation and schedule priority.

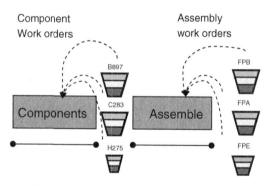

Finished Item Planner Bench

Part	Open Supply	On-hand	Demand	Available Stock	Recommended Supply Qty	Action
FPB	5453	4012	3200	60% YELLOW	6500	Create Work Order
FPE	3358	4054	540	66% YELLOW	3128	Create Work Order
FPA	530	3721	213	67% YELLOW	2162	Create Work Order

Intermediate Component Planner Bench

Part	Open Supply	On-hand	Demand	Available Stock	Recommended Supply Qty	Action
B987	16359	12000	8000	58% YELLOW	60	Create Work Order
C283	900	5532	960	72% YELLOW	30	Create Work Order
H275	1530	3721	713	74% YELLOW	15	Create Work Order

Figure 11.36 Work Order Generation from Strategic Stock Buffers

Stock Buffers' Role in Execution

The third implication that strategic stock buffers have on scheduling is the ability to match changing priority sequence to open work orders during the execution cycle time. In a dynamic environment, variation (demand, quality, process, etc.) will continue to affect the on-hand inventory position of these stock buffers over the execution timeframe. Figure 11.37 demonstrates how an expedite status or priority change is triggered during work order execution if a strategic stock item penetrates the red critical zone. The work order's percentage of buffer remaining determines work order priority if multiple parts enter the red critical zone. Color provides a general reference while percentage remaining provides a more precise priority reference.

Priorities can be changed to prevent a stock-out from passing variation forward and delaying the assembly release schedule, a missed shipment schedule, and a resulting service level stock-out at a distribution center. Managing the events in the tails blocks negative variation from spiraling into the supply chain and destabilizing the system and market performance. The ability to identify negative events and take action before they cause disruption dampens the bullwhip effect. When tied together through an entire supply chain and applied consistently between process units, as was the case at Letourneau Technologies (referenced in the Foreword of this book), the gains in flow are tremendous.

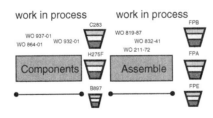

Component Items			
Order #	Due Date	Item #	Buffer Status
WO 932-01	5/19	B897	20% RED
WO 864-01	5/18	H275	25% RED
WO 973-01	5/20	C283	40% YELLOW

Finished Items			
Order #	Due Date	Item #	Buffer Status
WO 819 -87	54/24	FPB	19% RED
WO 832 -41	5/22	FPE	22% RED
WO 211 -72	5/22	FPA	43% YELLOW

Figure 11.37 Real-Time Execution Priority for Work-in-Process

One side of the Paretian tail (green and over top of green zone) shows where there is currently lost or hidden opportunity. The other side (red critical and stock-out) pinpoints where to focus and act to block a negative event from spiraling into an unnecessary expenditure, investment, and/or a market loss. Smart metrics are focused on the opportunity to identify butterfly levers in their early stages so they can be stopped from spiraling into extreme outcomes or disaster. At best, the center of a distribution can capture the past. It is the tails of the distribution where innovations occur and smart metrics are focused. The tails of strategic buffers safely allow companies to operate at the edge of chaos, identify issues, focus resources, learn and emerge the system to a higher level of order. Strategic buffers provide an early warning system as well as a much safer zone to learn and innovate within.

Once a higher system performance emerges, the other side of the buffer tail (green and over top of green zone) signals the opportunity for a focused investment reduction and a new benchmark is set to the improvement process. If the strategic buffers are stable, the system is successfully decoupled from the effects of dependent variation between those buffers and the result is shorter, stable, and reliable planning and execution horizons. If they are compromised the market lead-time strategy is compromised and market service levels become unreliable. Chaos and operating expenses spiral out from the loss of stability. It should start to be clear why and how smart metric objectives are focused on reliability, stability, speed/velocity, and improvement opportunity, and that buffer management is the smart metric performance assessment tool.

Control Points and Resource Scheduling

In our example, the first objective of component scheduling is to ensure the strategic control points (lathes and weld) can reliably execute the work released to them and deliver on time to the buffers they feed. The second objective is to ensure the alignment and attention of all feeding resources to the control points' schedules. Control points serve the following important functions and define the system's relevant information for all other resource schedules:

▲ As Pacesetters they set the cadence of the entire system.
▲ They are finitely scheduled in order to reliably meet customer and replenishment (strategic stock buffers) demand.

▲ Their finite schedule determines all other resource schedules and priorities.

▲ They are used to determine the levels of capacity investment (resource, labor and stock) necessary to deliver the market lead-time strategy and product volume plan.

▲ They set the relevant range of capacity and define relevant information for CVP analysis. The rate they generate throughput dollars, the cost, and the time frame to increase their capacity determine the system's relevant range of product volume, net profit opportunity, and fixed-cost investment.

Figure 11.38 demonstrates how components and assembly are scheduled. Both weld and lathe schedule priorities are pulled from the zone status of the strategic component buffers they feed. Strategic control point daily resource loading is choked to the total available finite capacity of each control point resource. The control points' scheduled capacity bar charts do not exceed the practical capacity of their resources. All other resource capacity schedules are allowed to float to the level demanded by the control points they feed because they all should have

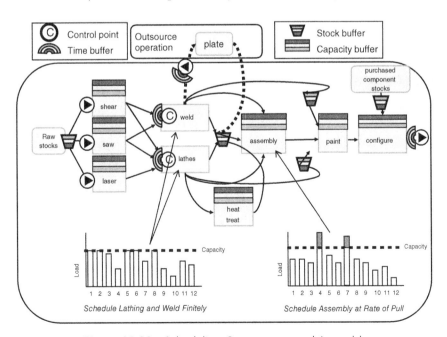

Figure 11.38 Scheduling Components and Assembly

sufficient capacity buffers to recover from temporary surges. The bar graph in the middle of the feeding resources shows temporary overloads and the recovery pattern.

Assembly is not a control point. They are scheduled at the rate the strategic stock buffers are pulling through to meet direct-ship customer orders. All resources in assembly have surge capacity beyond components' ability to deliver to the stock buffers they pull from. Assembly's capacity and the strategic stock buffers at each distribution center have been sized to absorb the normal demand variation the system experiences.

Time Buffers and Reliable Schedule Execution

Figure 11.39 demonstrates a Paretian view of the time protection designed to break variation from impacting a critical due date. The due date can be a scheduled time at a control point for a work order, a scheduled time to transfer work to an outsource vendor, a due date for delivery to a customer, or a strategic stock buffer. Because we are measuring time, we are either on time, early or we are late. The objective of time buffers is to ensure schedule reliability at critical points and system schedule stability. Time buffers *are* smart metrics in action. The same intuitive color coding exists for time buffers as stock buffers. The color zones indicate the signal strength and priority of work for the feeding resources. A status check is made on work orders in the red zone that have not arrived at the control point and a determination is made if any action is required to ensure their arrival. The red zone

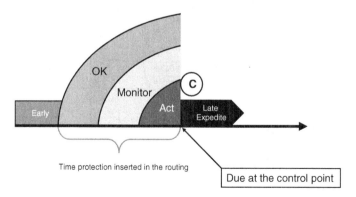

Figure 11.39 Color-Coded Zones Indicate the Time Remaining for Work to Arrive on Schedule

is the decision zone—act or don't act. Expedites are disruptive and if the control point is well protected with work and the component stock buffer is in good shape, it is up to the buffer manager to make the call if action is truly needed. If the opposite state exists they will initiate whatever action is appropriate. Work orders in the late zone trigger an expedite if an expedite status was not previously assigned in the red zone.

Time Buffers as the Execution Feedback Loop for Manufacturing

Focus is of course on the tails of the distribution to provide the relevant information for execution priority. Figure 11.40 demonstrates the dynamic nature of time buffers. The depth of the buffer zone penetration is tracked in real-time and moves dynamically across the zones. When a work order arrives, a transaction is required to note its arrival and record the zone status it arrived in.

When something is yet to be received, the clock is still ticking on the time it has remaining to meet the control point schedule. This is indicated by the arrow penetrating the buffer zones. The net amount of positive and negative variation a work order experiences en route determines what color zone it will arrive in the buffer. In Fig. 11.40, you can see the first work order arrived in the red zone, the second arrived early, the third in green, and the fourth in yellow, and so on. If a work order arrives anytime in the red zone, it will still be on time to the control point schedule. The red zone has two objectives: provide an early action warning for work in danger of not arriving on time to schedule, and provide a mechanism to capture the variation event and its cause for later trend analysis. A zone status of early, red, or late requires a reason code entry. The capability for feeding resources to attach a note electronically to the work order and record the cause of the buffer penetration provides the necessary information for future process improvement. Identifying the event at the point of occurrence ensures valid reason code identification and meaningful trend reporting and analysis.

All of the work orders in this example have arrived on time and the buffer has absorbed all of the process variation experienced at prior resources. The current buffer size is ideal for both red zone objectives. The schedule is on time and there is an entry in the red zone to capture and

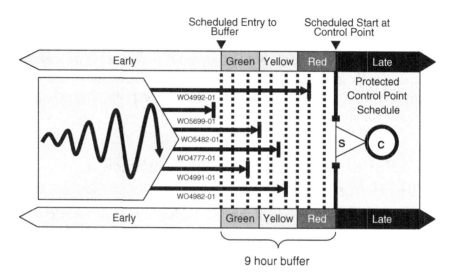

Figure 11.40 The Time Buffer Breaks Variation When the Work Arrives On Time

focus future improvement opportunities. There is also an arrival in the early zone. Tracking the cause of early arrivals provides the opportunity to correct sloppy time or routing standards. This tightens the schedules, allowing a later release and improved future velocity and flow of that part.

Time Buffers Require Ten Zones

In order to provide clear visibility, direct proper corrective actions, and collect data about events, the buffers must be configured in a specific way. Time buffers account for the same time horizon from the two different perspectives and this requires the ten status zones display shown in Fig. 11.41. In our example company, the total time buffer for lathe is nine hours divided into three hours for the green, yellow, and red zones respectively. The top five zones monitor the performance of the feeding resources. When a work order is received, it is ready and available to the critical resource and it drops to the lower horizon in the same status zone it left.

There is only one buffer of time but two different views for two different purposes. The top zone is a visible status of the performance and priorities of feeding resources. The bottom zone provides a visible status of control

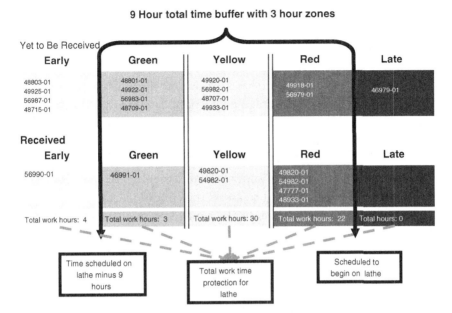

Figure 11.41 Buffer Boards Measure Feeding Resources Performance, the Flow Status of Individual Work Orders, and the Control Point's Schedule Status

point protection and performance to schedule. In Fig. 11.41 the "yet to be received" horizon has one late order. The manager responsible for this buffer should have triggered an expedite and priority actions should already be underway. The manager would also have reviewed the two red zone orders' statuses to determine if they require any action to arrive on time. The buffer managers, schedulers, and control point resource managers are the primary adaptive agents of the system.

The lower five zones monitor the depth of time protection of the work waiting for the critical resource. Time protection is determined by the total critical resource hours, of all work orders waiting for their time on the critical resource by zone. The total load hours in the late zone provide a clear picture of how well the control point is executing to schedule. It also provides an early warning when the work load provides insufficient protection and the control point is in danger of experiencing downtime from starvation. On average we would expect work orders to arrive in the yellow zone and an average workload protection of 4.5 hours per lathe. Again, it is all in the monitoring and measuring of the events in the tails.

In our company example in Fig. 11.41, the drum has only one late order and it has not arrived yet. There is a total of 59 hours of work load protection. The lathe operation has 12 lathes and 4.5 hours of work would be 54 hours of expected average protection. The medium lathe is the most constrained but can be offloaded by the large lathe and all three are control points. The scheduler's only caveat must be that the combined load of medium and large cannot exceed 90 percent of their total capacity. It should be left to the lathe supervisor to assign his or her resources. One buffer board is sufficient to assess the condition of the lathes. The feeding resources are on schedule, the control point buffer is stable, and the control point is on schedule with the work that is available for them to process. The system is stable and delivering reliably on schedule. To be truly effective, time buffer boards should be maintained as close to real-time as practical and be visible across the organization. The authors strongly discourage manual boards and cards. They are labor intensive, prone to error, and can be viewed only by standing in front of them. Visibility to the control points and decoupling points status is critical to maintaining system coherence.

Time Buffers During Work Order Execution

Time buffers are the visible and real-time feedback loops and the focal point of smart metrics to direct execution action and provide trend analysis information for focused improvement and investment. All buffers are chosen to strategically break dependent variation and protect a critical point or process or market commitment. All resource priorities are aligned based on work in the red and late zone of the buffer they feed. If and when there is resource or material contention at feeding or service resources (maintenance, quality, manufacturing, engineering), priority and the action decision is made based on the zone status. Visibility across the system provides all agents with the same relevant information and action priority.

Time buffers protect the scheduled start date of a work order at a control point in the supply chain process. Time buffers also protect a scarce capacity resource, shipping schedule, and ensure on-time transfer to an outsource operation. Time buffers are intended as shock absorbers and are sized based on the protective capacity of the feeding resources and their ability to surge and recover from normal dependent and process variation. The more surge capacity and the more reliable the feeding resource processes, the smaller the time buffer required. The time buffer is a strategically sized insertion of time in the routings for each part routed through a critical control point. Time standards for parts should only account for run time, setup time, cure time, or dry time. All padding is removed and accounted for in the time buffers where it has strategic value and can be monitored and measured. Parts with padded time standards or standards that are too small will routinely arrive in the early zone or the red and late respectively. They will be tagged for review with the appropriate reason code. All standards should be systematically corrected over a period of time. The reduction in variation from better standards often results in the ability to reduce the buffer size.

Time buffers provide protection from a lost-time event and create stable and reliable flow to the control point they protect. If strategic buffers are stable, the control points are stable. If control points are performing to plan, the system is performing to plan. The speed and velocity of synchronized flow to and through the control points determine the speed and velocity of flow through the entire supply chain.

Minimizing Control Point Disruption

There are four common causes of control point disruptions and all of the strategic buffers are used in some combination to minimize the negative effects, identify the root causes, and continuously work to remove them. The table below describes these disruptions and the corresponding ways to mitigate them.

Control Point Disruption	Minimization of the Disruption
Material, component parts unavailable to start date; end item product unavailable to ship	– Commit to maintaining investment in strategic stock throughout the system. – Check material availability prior to work release to operations. – Monitor stock buffer critical red and stock-outs and act. – Review red critical and stock-out trends and adjust part buffer profiles and zones.

(Continued)

Control Point Disruption	Minimization of the Disruption
Producing parts not required now	– Carefully control the release of material to coincide with the control point schedule.
	– Stagger the release of material only within the proper horizon—*not before.*
	– Monitor early receipts in the time buffers and enforce reason code reporting, Pareto analysis, and improvement trends.
	– Monitor strategic stock flow indexes and buffer trend reporting.
Blockages	– Commit to maintaining protective capacity to prevent feeding resource from being temporarily overwhelmed.
	– Monitor red zone and late receipts in the time buffers and enforce reason code reporting, Pareto analysis, and improvement trends.
Downtime (planned and unplanned)	– Combination of stock, surge capacity of feeding resources, and inserted time buffer in front of the control point.

Schedule Execution Coherence and Priority Alignment

Figure 11.42 uses our Company CMG example to demonstrate where each resource in component manufacturing is focused for its work priority. Weld, lathe, and the outsource plate time buffers are the priority management indicators for shear, saw, and laser activity. Each of these resources has the same measure and performance criteria—on-time delivery to the buffer they feed. The control points' performance measurement is on-time delivery to the component stock buffers they replenish. Component stock buffer statuses direct the control point priorities when they are in danger of stocking out.

Figure 11.43 demonstrates each resource in assembly is focused on the on-time delivery to the time buffer in front of shipping/end item warehouse. This time buffer is a barometer of the state of on-time delivery to both the customer order date and the buffered stock due date. The end item stock buffer statuses direct and align all three resource priorities when they are in danger of stocking out. Time buffer board visibility provides

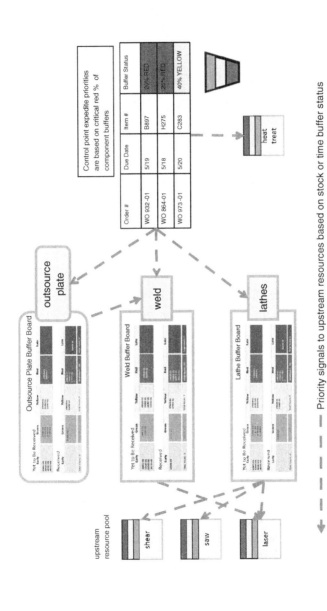

Figure 11.42 Execution Performance Measurement and Priority Alignment for All Resources Feeding the Component Stock Buffers

269

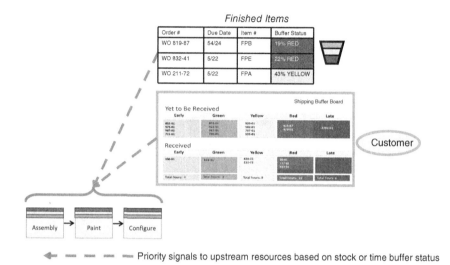

Figure 11.43   Execution Performance Measurement and Priority Alignment for Assembly

drill-down capability to individual work orders' status. Sales access to buffer boards allows them to review late orders and make decisions on individual orders based on the customer priorities as well as keep their customers updated on individual order status. Not every order has the same criticality to the customer. Visibility and real-time status allow sales and operations to make better decisions together when there is a potential conflict because they have the same view of the system and the relevant information to understand the impact of the expedite actions.

Assigning Roles and Responsibilities

Component and end item inventory buffer managers ensure components and products are available to meet the assembly schedule and shipment due dates with the minimum investment in working capital and expedites. Their duties might include:

▲ Allocate product inventory to sales orders for shipment, components to assembly orders, and materials to component manufacturing scheduling when there is material contention.

▲ Monitor inventory levels and initiate a reorder.

- Monitor inventory in red critical or below and initiate an expedite status for the work orders in process to the responsible control point buffer manager.
- Monitor buffer sizes over time to adjust buffer profiles and zones.
- Monitor and trigger planned adjustments for projected sales trends.

Materials inventory buffer managers ensure material availability to meet control point scheduled start dates with the minimum inventory investment. Their duties might include:

- Work with the component and end item buffer managers to allocate materials for work order scheduling when there is material contention.
- Monitor inventory levels and initiate a purchase reorder.
- Monitor inventory in red critical or below and communicate the change in priority or expedite status to the vendor.
- Update purchase part order receipt dates with changes in material availability.
- Monitor buffer sizes over time to adjust buffer profiles and zones.
- Monitor and trigger planned adjustments for projected sales trends in conjunction with end item and component inventory buffer managers.

Control point buffer managers ensure the feeding resources are capable and able to deliver to the buffer on time to the control point schedule. Their duties might include:

- Ensure work orders are received on time in the buffer.
- Ensure buffer transactions are timely.
- Reconcile changes to the control point schedule to the work that is available to process.
- Initiate and communicate priority changes to feeding resources and service resources.
- Initiate and communicate expedites to feeding resources and service resources.
- Work with all resource managers to determine how (what actions) and who (resources involved) will execute the expedite recovery plan.
- Determine when, who, and how much overtime is required at all feeding resources based on both the amount of work load protection available to the control point and the status of late and red work orders yet to be received.

Control point managers ensure both scarce capacity exploitation and control point schedule compliance. Their duties might include:

▲ Determine the best use of scarce control point capacity to the work order priority from the total available work load.

▲ Work with and through their buffer manager when available work in the buffer falls below a safe protection zone.

▲ Work with and through the buffer manager when offloading is determined to be necessary and initiate overtime when the backlog of available work in the buffer is growing and/or late.

▲ When backlog occurs that cannot be worked out with overtime or offloading, inform the scheduling to choke the rate of work released to the feeding resources.

Schedulers ensure market-driven realistic and reliable schedules that exploit scarce capacity resources. Their duties might include:

▲ Finitely schedule key production areas to meet real demand requirements.

▲ Generate capacity tested due dates based on the priority of the inventory buffer status and/or customer request date.

▲ Work with Sales for specific work order and customer priorities when short-term demand spikes create short-term capacity contention.

▲ Determine when materials should be released to arrive at the control point schedule date.

▲ Determine if materials will be available prior to releasing orders.

▲ Monitor:
  ▼ The control point status protective capacity and late backlog and take actions to choke work order release when necessary.
  ▼ Amount of unreleased late work and its impact on market lead times.

▲ Work with Sales to determine if the expanded lead time will impact the market strategy.

▲ Initiate planned overtime and/or outsourcing actions when the backlog of unreleased orders is growing beyond the short-term surge capacity to recover.

Sales ensures the product and customers are aligned to the highest strategic contribution for total Return on Investment. Their duties might include:

▲ Work with the end item inventory managers to provide the timing of all known and planned changes in product ADU (seasonality, promotions, product introduction and obsolescence).

▲ Identify premium pricing opportunities for strategically reduced lead times is captured.

▲ Work with operations to prioritize by customer and order when capacity contention exists

▲ Communicate with the customer on order status that will be different from the customer request date.

Engineering, Manufacturing Engineering, Quality, and Maintenance ensure their priorities are based on the positive impact on the scarce capacity resources and their buffers.

Larger, More Complex, Multisite System Control Requirements

Larger and more complex organizations need a central strategic planning and logistics function to de-conflict resource contention between brands or business groups. This function's job is to ensure smooth flow and maximum strategic contribution within the company. Below is a list of duties they should handle:

▲ Provide weekly reports to all business heads.
  ▼ Coordinate efforts, where needed, to improve system stability.
  ▼ Review control point/shipping buffers.
  ▼ Review control point utilization and compare to demand.
▲ Provide monthly reports on stock-outs and turns to all business heads.
▲ Scarce capacity reconciliation
  ▼ Report on and review each critical area's planned time standards to actual performance.
  ▼ Resolve scarce capacity/schedule conflicts that affect multiple business groups/brands.
▲ Central buffer management
  ▼ Work with individual business groups/brands on end item buffer levels.
  ▼ Adjust levels based on demand requirements/trends and capacity requirement trends.
▲ Strategic contribution and finance
  ▼ Recommend the best balance of service performance and the sales opportunity for selling extra capacity after the supply chain needs

are met. For vertically integrated supply chains, examples would be logs, steel, foundry castings, and so on.

▼ Provide business groups with strategic contribution and through-put dollar velocity rates per unit of scarce resource for all end item products with true constraints.

▲ Review trend analysis of all buffers and reason codes and determine project priorities and coordinate assignment of process improvement actions and projects.

▼ Coordinate engineering changes and priorities.

▼ Monitor nonreplenishment inventory and make recommendations for the disposition of excess.

▼ Outsourcing and all make/buy decisions.

▼ Review capital expenditures.

▲ IT liaison for all software issues relating to manufacturing, distribution, and process improvement projects.

The measurements are not part of the IT liaison. Measurements for this central function should include:

1. On-time performance of the entire system
2. Replenishment inventory performance
   a. Stock-outs
   b. Flow index
3. Throughput dollar velocity per true capacity constrained control points

The Cheat Sheet for a Demand Driven Performance System

Ask and answer five key questions hourly, daily, weekly, and monthly.

Employing an effective feedback and accountability system requires answering these questions daily, weekly, monthly, and quarterly at different levels of the organization. The point is to get an operational system in place quickly that can identify and break variation, and a feedback system that can direct the process of execution, feedback, and ongoing improvement.

1. What is the state of the control point(s) and the buffers they feed? Review the time buffer work order status of red zone and late zones in both yet

to be received and received. Hours of work load protection available to the control point. Stock buffers in the critical red or stocked out.

2. Is the state trending worse or better? Track buffer protection continuing to erode; growing red and late zones or is the buffer protection rebuilding and control points back on schedule? Act or don't act?

3. If the trending is worse, what is the recovery plan? Locate the work orders with the highest buffer penetration and direct the right resources to get them back on track. Maintenance, quality, manufacturing engineering, overtime, and so on are directed to the specific area in trouble causing the late work.

4. Was the recovery plan effective? Did the state of the buffers and the control point(s) improve?

5. What preventive measures are in place to keep the root issue from recurring? The reason code data of work entering in the time buffer in the red, late, and early zones is analyzed and used to direct process improvement and investment focus.

Ensure the five critical data capture points to provide the smart metric answers are timely.

Work flow data capture points must be recorded in order to monitor, control, and provide information for all four execution smart metric objectives for the system. Answering the five key buffer management questions requires the following minimum transaction points:

1. On time to release measures whether scheduling is releasing jobs when they should. Orders released late endanger their respective time buffer.

2. Work order actual start time at the first operation. Delay between planned start time and actual start time is buffer penetration time. Negative variation at the starting resource provides an early warning of potential buffer penetration. Time variation greater than 35 percent of the time buffer zone should trigger the starting resource to record the cause of the delay on the work order to provide useful buffer board reason codes for a red or late arrival in the buffer.

3. Entry into every buffer/time and stock. Real-time, visible buffers are only useful if the transactions at each and every strategic buffer are timely. Late transactions send a false signal to act. Time buffer receipt transactions require a reason code assignment for work entering early, red, or late.

4. Control point start time transactions. Starting a work order removes it from the control point buffer board and updates the total protection load hours displayed by zone. The control point schedule is the system schedule. The received zones of the buffer board are the barometer of both the control point's current on-time to schedule and its ability to stay on schedule.

5. Control point stop time transactions. Once again, the control point schedule status is the system schedule status. Negative variation of completion time at the control point provides an early warning of the buffer penetration at the stock or shipping time schedule buffer. Time standards at the control points are critical because they govern the output of the system, are used to judge strategic product contribution rates, and determine the system schedule. Sloppy standards for control points create an unreliable and unstable schedule and unreliable information for decision-making. Control point standard deviations are reviewed and corrected as part of the regular improvement feedback cycle.

Plan an executable schedule focused on system flow and the market lead-time strategy.

1. Control points are loaded to practical finite capacities (for example 90 percent of total available capacity). In addition surge capacity should be set aside for planned demand drop-ins if they are part of the agreed-to market strategy. This means that practical finite capacity load should be scheduled less than 90 percent. Reserving ten percent for drop-ins would reduce the scheduled load to 80 percent. Examples of relevant drop-in business would be service parts with high contribution or premium priced expedites that are strategic market factors (e.g., machine or aircraft down).

2. Schedule priority is determined first by:
   a. Stock buffers with critical zone penetrations (greater than 50 percent of yellow) and then by customer promise due date in companies with both make-to-stock and make-to-order. *Yes* stock buffers can take precedence because they protect multiple customer orders.
   b. All remaining stock buffers with yellow zone penetration.

3. Massage the schedule. When more capacity is available, scarce capacity can be safely stored:

   a. Increase the order quantity of parts previously scheduled by X% of their ADU. Parts with high setups and good flow indexes are the *only* candidates to consider.

   b. Schedule any remaining available capacity of scarce resource control points by green zone penetration to parts' with low control point setup times.

4. Work orders are *only* released to operations if material is available.

5. Control point execution is monitored and the work order schedule and release adjusted accordingly:

   a. Work orders are choked to the pace the control point is performing. In Fig. 11.41, the lathe time buffer top of green zone is 9 hours of protection or 108 hours of work load for all 12 lathes. If the received work load exceeds 108 hours, it is time to choke the schedule until the imbalance is corrected.

   b. As variation is reduced, expect control point gains in velocity and increased available capacity to schedule. The 10 percent set aside for variation can be incrementally reduced over time. The drop-in capacity reservation is maintained in line with the market opportunities. It is part of the process improvement cycle—*watch for it and track it.*

Expect reliability, stability, and speed.

The system will perform reliably because:

1. All localities are performing work in the same prioritized sequence directed by a buffer status that is a direct reflection of shipping schedules and end item stock buffer priorities. The status of each buffer and the control points are visible to all decision makers and everyone has the same view of the system priorities.

2. Localities are encouraged to make or buy only what is necessary relative to true market pull:

   - Individual resource utilization is not the responsibility or a performance measure of the resource manager.

   - The shop is scheduled/choked to the rate of the control points or market pull from the end item stock buffers.

3. Real time visibility and priority alignment means little to no conflict regarding what is the right action because everyone has the same view of the system, the same measure, and the same view of progress check points (the buffers).
4. People will be more inclined to take responsibility and be proactive when:
   - They are not trying to satisfy conflicting measures or local resource subgoals.
   - There are clear lines of authority and responsibility.
   - The effects of their actions or inaction are clearly visible and so is everyone else's.

The system will remain stable because:

1. Time buffers block variation from being passed to the control points.
2. There is an executable schedule for both control points (finitely scheduled), and all other resources have sufficient surge capacity to recover buffer erosion.
3. Stock buffers decouple dependent variation at critical points in the supply and distribution network and the product structure.

The system flow speed/velocity will be maximized because:

1. All resources are focused on protecting the control points' speed and velocity.
2. Real-time visibility and priority alignment means the right work is passing through the system based on the market pull priority.
3. Resource areas performing below scheduled standard time will deliver in the red zone and the event cause recorded. Reason code trending will target the resource area and the cause as a high-flow process improvement opportunity.
4. Resources' key performance metric is on-time delivery to the buffer they feed.

The targeted system improvements will increase ROI because:

1. Good reason codes targeting root cause are collected for events arriving in early, red, and late.
2. Data from both tails of all of the buffers are trended for weekly and monthly review.

3. Specific improvement tasks are assigned to appropriate individuals.
   a. Correcting time standards is assigned as a routine job task of manufacturing engineering.
   b. Correcting stock buffer profiles is assigned as a routine job task of the buyers and planners responsible for the strategic stock buffers.
   c. Improvement projects are limited to the very few top priorities. Limit multitasking and focus managers on flow of high-value improvement projects for the system.
4. Weekly and monthly reviews of the improvement projects, progress, and results is monitored and confirmed by the improvement in the buffer trending.

Remove cost-centric competing measures.

The current world of supply chain performance measurement systems is irreparably broken and must be replaced. Innovating on top of flawed rules, bad math, and conflicting metrics is the cause of the tremendous conflict, variation, and waste. Existing cost-centric measures cannot be augmented; they must be replaced. Global supply chain companies are already attempting to straddle two worlds and a foot in both is a recipe for mediocrity at best and extinction at worst. Financial accounting measurements occur in the accounting cycle monthly, quarterly, and annually. A supply chain performance measurement system is a planning and execution feedback loop with no endpoint. It is continuous, interactive, and dynamic. It is based on time metrics designed to synchronize the right priorities and drive flow, identify and remove variation, and focus on strategic contribution and sales opportunities. The two systems cannot cross over; they have different purposes and opposing rules.

Identify all competing cost-centric metrics in your strategy session and remove all of them prior to going live with a new performance measurement system, or you will simply end up in the same conflicted space again. Smart metrics promote actions that benefit the system as a whole. Bad metrics promote actions that benefit a local area to the detriment of the system as a whole. Their actions are almost always in conflict. Identify and eliminate all conflicts always.

CHAPTER 12

Summary

"The greatest obstacle to discovering the shape of the earth, the continents, and the ocean was not ignorance but the illusion of knowledge."

—Daniel Boorstin

In our introduction, we presented the idea of a deep truth as an underlying core belief that lies at the heart of how an individual, a company, a culture, and even the world see reality. It is something so universally accepted it is not even questioned. Our deep truths are simply the ground we stand on and the window we look through. They determine our viewpoint of the world and how strongly we feel about what we see. The authors have made their case and are asking you to challenge a very deep truth:

Today's Deep Truth:

Decreasing Unit Cost = Increasing Return on Investment

What did we learn about this deep truth?

▲ The whole idea of a least unit product cost is simply bad math—an inappropriate use of an equation that both economics and even physics reject.

▲ Legislation created a reporting requirement that has become the focus of accounting information and replaced, almost by accident, the real definition and rules for relevant information for decision-making and product costing.

▲ All of our information systems are hard-coded and/or configured to compile cost reporting and resource area measures from the wrong or misapplied rules and assumptions about how costs and revenue behave.

▲ Unit cost has become such a deep truth that an entire discipline about what defines relevant information has been all but lost.

▲ Even those who know what relevant costs should be operated inside a system that is not capable of providing relevant information in a relevant time frame in which to act.

▲ People no longer even question taking actions they know will lead to predictable and dire negative consequences that they must deal with later.

Did we make a compelling case that today's Deep Truth is totally, completely, and unequivocally false? It is the hallmark of any deep truth that its negation is also a deep truth.[1]

Performance Measurement Built on a New Deep Truth

In today's globally competitive environment, new decision-making tools are required to monitor, measure, and improve the business based on the reality that it is a complex adaptive system. A demand driven information system is designed to plan, execute, and focus/prioritize improvement using a real-time feedback loop focused on the flow to and through strategic control points and decoupling points. This aligns all to the system view and strategy. The points for measurement and real-time feedback are relatively few and strategically chosen to protect critical resources and/or hand-offs between processes or subsystems. These strategic buffers of stored time are sized to break dependent variation and provide visibility and synchronization to resource managers so they can act. Buffers provide all of the information needed to judge the state of the entire chain and direct attention or action as well as focus opportunities for improvement and investment.

A measure is simply a reading at any point in time of the current state of the system relative to the standard or plan the system was directed to execute. The information is not used to reward or punish individuals but it is used to measure performance of both the system and the individual resources' ability to protect and improve system flow. Real-time exception

---

[1]Schilpp, P.A., ed., "Discussion with Einstein on Epistemological Problems in Atomic Physics," in *Albert Einstein: Philosopher-Scientist*, 1949, page 240.

feedback is needed to identify issues and their root causes proactively so people can take timely, appropriate action. They are also trended over time to provide focus to permanently remove recurring issues or events that routinely block flow.

The Need for Steady Feedback

The only certainty facing most organizations is that conditions do not stay the same. A supply chain performance measurement system needs a steady feedback of system performance and regular adjustment of the actions of different resources to stay synchronized. A shift in a supply or demand condition will result in the need to modify activity, sometimes dramatically. The longer the lag between the change signal and the corrective action, the more dramatic the course correction will be and the more protection and resilience in the form of time protection is required. Without an effective feedback mechanism, people will continue to drive toward their planned target without recognition that the conditions have changed. In other words, it will result in actions that are believed to be the right thing for improving ROI but can actually be hurting the company. In a dynamic interdependent system, the same actions may have been absolutely the right thing to do yesterday but are dead wrong today. Everyone throughout the organization must understand that the feedback mechanism drives measurement away from a fixed tactic or action and toward the action necessary in this moment.

People can sometimes have a problem understanding that the target can stay constant, but the means to achieve it, and therefore the measurement, may change. Buffer management clearly makes this connection for people with simple, visible signals. It creates focus and directs action. Although on-time delivery to the buffer is always the target and the measure, very different actions are needed every day, dependent on the real-time state of all buffers (time, stock, and capacity) across the supply chain. Decisions include when or where to flex labor; batch setups; where to direct maintenance, quality, or engineering; and which distribution center needs what product first. Efforts change according to the status and cause of the buffer penetration. An effective operational planning and control solution is a prerequisite to a functional (rather than dysfunctional) performance measurement system. The operational system will fail to properly execute or sustain without an effective way to provide feedback on the current system status that synchronizes and de-conflicts decisions and actions every hour of every day.

Focus on Improvement

In contrast to the fixed target of individual KPIs (quality, inventory, cost, sales, and service), a feedback system does not have an endpoint, but provides continual monitoring of flow to determine exceptions or issues. The regions in buffer management used to monitor flow are sized to allow response to an exception and react quickly enough to maintain the desired flow. The red zone is designed to be the edge of chaos and provide a safe learning ground to improve. Additionally, by conducting an analysis of these exceptions and identifying and eliminating the causes of exceptions, a process of ongoing improvement is achieved. A rigorous process of identifying, analyzing, learning, and improving the system is the only way to tie process improvement to the system ROI.

By definition, an effective performance measurement system must consider all of the tactical objectives that determine ROI simultaneously rather than any single KPI performance standard. Understanding the interdependent nature of the tactical objectives is a necessity in the performance-feedback system. Companies that understand the above and incorporate it into their financial, strategic investment, and asset-performance analysis thrive and continue to grow regardless of the economic circumstances in which they are competing. Those who do not understand the significance of strategic buffer management and its role in continuous improvement will commonly see any improvement stagnate and decline after experiencing initial "brilliant" results. They are repeating the past actions that resulted in their success but their reality has moved on. They will inevitably end up retrenching to their previous level of performance. The authors have witnessed this repeatedly when companies change ownership, executive leadership, and/or information technology.

If a manager gains visibility—through buffer management or any other mechanism—to a potential problem in advance of it affecting performance, this is a great sign that the system is working. Do not mistake this sign with an absence of issues. Companies and the humans who work in them have no shortage of issues—the system is dynamic by definition. The ability to separate the issues from the noise and act on them before they affect the company's flow performance is powerful but it only takes you so far. If issues are captured, clarified, understood, and resolved for the future, then bottom-line performance improves with the same or less capital investment.

Each issue/problem identified and understood is an opportunity for improvement. Unfortunately many individuals of a fixed mindset view the presentation of the problem as an indication that the system is *not* working. To accept that identification of potential problems is vital to the measurement and execution system working effectively is to accept full responsibility and accountability at all levels of the organization. This thinking is not entirely comfortable for everyone. Without top management understanding and owning this view of the system, there is very little hope that the rest of the organization's management will be able to adopt the right mindset.

When individuals in an organization are not synchronized around the right action to take, by definition, they are wasting the capacity of their resources. Conflict over direction and the right action inhibits the exploitation of flow to and through the control points and casts an organization-wide doubt in leadership's knowledge of both process and logistical flow as well as their ability to manage. Leadership needs to understand how buffer management works and use the five management questions presented at the end of the Chap. 11 to direct their monthly review with their management team. When the primary day-to-day measurement of the state of flow and additional opportunity of the system is directed by buffer management, then weekly, monthly, and quarterly reviews must incorporate the information as well. Most importantly, the entire organization must understand that execution measurements are synchronized from local to global through measuring the resources of each feeding link to its strategic buffer. Every improvement in flow or reduction in variation allows for a reduction in the strategic buffers supporting the feeding links while maintaining service levels. It's simply good math, statistical analysis and common sense.

The problem with fixed targets, standards, and metrics is they are perceived and often assumed to have an endpoint—a state of "being achieved"—and therefore do not promote ongoing improvement. The first step is to tear down the walls that separated the organization into different business units or functions. It is a necessary condition to align strategy, signals, and action around a demand driven flow map. The second step is to strip out the performance-management systems and metrics that encouraged the whole organization to be viewed as the sum of its parts. This local viewpoint is the source of conflict over the effective use of shared resources (i.e., capacity, inventory, and so on) and tactical priorities.

The most important thing to remember for all individuals playing a role in a demand driven performance measurement system is that this is a *thinking and evolving* system, not a *fire and forget* system. Fixed metrics will often point toward a single direction that, regardless of the need of the system, will continue to motivate efforts independently toward its achievement. A truly effective performance measurement system will point everyone in the direction that will have the greatest return which, by definition, is a growth model. The buffer management feedback system provides the relevant information to make day-to-day decisions in line with the organizational ROI measure. As changes occur, people must think, adjust, and adapt to achieve the greatest potential of the organization.

It is impossible to separate the smart metrics from the information feedback loop in a demand driven performance system because the feedback loop *is* the decision-making and the measurement tool set. The visible statuses of both the buffers and the control points are the feedback loop on the state of the system. The key to smart metrics is to generate the expectation that identifying, acting, and learning from the issue is part of everyone's job description and, responsibility. The focus is to understand the interdependencies and act in concert together, not against each other. The dashboard of strategic control points and decoupling points is directly connected to cash flow and ROI opportunity. The performance standard is always viewed in relationship to time and cash flow, not to unit cost.

A Final Thought on Information Technology

There are two major information technology stumbling blocks to creating a demand driven performance system. The first problem is again ontological: what type of reality do the major software providers assume supply chains are attempting to control and manage? In the authors' experience, working with dozens of different ERP systems, today's ERP, MRP, and DRP systems and modules are stuck in a linear world and a cost-centric strategy. The proof of this incredible mismatch between the need to plan, schedule, and deliver flow and the software tools available in current ERP systems was presented in Fig. 1.1, "Companies using Spreadsheets to augment MRP." In a nutshell, does big software even remotely understand the nature of the problem? Based on experience with and personal knowledge of the inner workings of large software providers, the authors say the answer is "no," or, at best, "not yet."

The second stumbling block has to do with the role that IT is taking in most organizations. The authors sense a huge undercurrent of frustration within companies. Operations personnel are frustrated at the lack of solutions made available to them to combat the volatility and complexity they are experiencing. From many perspectives they see IT closing doors to innovation and improvement rather than opening them. On the other hand, IT has become extremely frustrated with Operations for working around and outside of the system. On several occasions the authors have been present at large executive meetings where tremendous results have been presented concerning the use of new methods and tools. Despite creating huge and exciting business results, IT's response has often been contained to concerns about the lack of compliance to the existing IT infrastructure. "Fine, you made $5 million in three months, but you did it outside the system. You need to do your solution inside of our system." And when informed that the system is incapable of supporting the solution? "That can't be, we have one of the best ERP products on the market. It supports all the best practices out there."

Companies spend a fortune on technology and devote tremendous time to install it. After this effort and expense, top management, guided by IT, understandably wants everyone to use it in a uniform manner. This can be problematic when considering the first stumbling block above. Operations and Distribution are being asked to work in a manner that supports the software rather than have the software work in a manner that supports operation and distribution flow and the strategic market objectives of the company. Decades ago, IT generally reported to finance and a logical case could be made based on cash flow and operational needs. The CFO who understood their operations and grew up in their plants immediately grasped the ideas and the need for change. In the New Normal, information technology is both core to the business and increasingly complex. IT now reports directly to the CEO or COO and in many instances IT appears to be wholly disconnected from the functions they provide service to and the strategic objectives of the business.

Operations and IT seem to be diverging at an alarming rate. We have observed IT organizations that are completely incapable of understanding and orienting around the business needs and the idea of flow. We have observed that operations people have little knowledge and regard for the needs and objectives of the IT organization. Both sides simply don't know what they don't know. The solution will require convergence rather than continued divergence. The minimum necessary conditions outlined in this book for a supply chain performance management system and smart metrics undoubtedly will require changes in the existing information technology features and functions in most ERP systems if that convergence is to happen. This subject will have to wait until the next demand driven book.

Once a deep truth is negated and replaced with a new reality view, there is no ability to buy-in to the old deep truth view. You can't unknow what you know. For the people who have taken this journey and succeeded, it is incredibly frustrating, difficult, and unfulfilling to find themselves working in the old paradigm again. Now that you know how deep the rabbit hole goes, there are three potential courses of reaction or action you can experience after closing the book:

1. Adopt the Calvin and Hobbes motto of "ignorance is bliss," cover your hands over your eyes, and ride the little red wagon over the cliff.
2. Jump up and yell "I know all of this and no one will listen!" Then continue to blame the system you are an integral part of as it races the red wagon toward the cliff.
3. Feel affirmed that you "knew" there was a smarter way and let curiosity lead you there. Learn more and involve the people you work with to learn and challenge their assumptions along with you. The authors' experience is people are smart, care deeply about the red wagon they are riding in, and are looking for common sense leadership.

No one likes taking actions they know to be "wrong," even when they see no other viable alternative.

More on the Strategic Replenishment Buffers of Demand Driven MRP (DDMRP)

This appendix will serve as a supplement about how DDMRP buffers are sized and how they are replenished. Chapters 23–28 in the third edition of *Orlicky's Material Requirements Planning* by Carol Ptak and Chad Smith extensively explain DDMRP and its buffers. For more information including white papers, podcasts, and videos, please go to www.demanddrivenmrp.com.

Additionally, the International Supply Chain Education Alliance offers an internationally accredited and available certificate program on DDMRP called the Certified Demand Driven Planner (CDDP) program. You can learn more about the CDDP program at www.demanddriveninstitute.com.

Sizing Buffers

Figure A.1 shows the components of sizing DDMRP buffer zones. The green zone is the heart of the supply order generation process embedded in the buffer. It determines average order frequency and typical order size. It is sized either by the part's minimum order quantity (MOQ), minimum order cycle or a percentage of the calculated average daily usage (ADU) multiplied by lead time, whichever results in a greater number. The ADU is typically calculated through the use of a rolling horizon looking backwards (e.g., the past 90 days). This percentage of ADU is determined by the lead-time category that the part falls within. The longer the part lead time, the smaller the percentage of ADU used in the equation. This is meant to force more

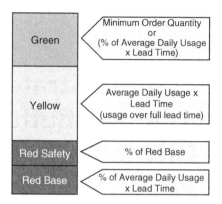

Figure A.1 Strategic Replenishment Buffer Sizing Factors

frequent orders for long lead-time parts, typically as frequently as order minimums will allow. This creates a conveyor belt effect as much as possible for long lead-time items.

The yellow zone is the heart of the coverage and shock absorption embedded in the buffer. As such, the yellow zone is always set to 100 percent of ADU over lead time.

The red zone is the risk mitigation embedded in the buffer. It is a summation of two separate equations. The first equation establishes a base level (red base in Fig. A.1) using a percentage of average daily usage multiplied by the part lead time. The next equation multiplies the previous equation's output by a certain percentage. This percentage is determined by the variability category that the part is placed in. The higher the variability, the higher the percentage used.

Figure A.2 is an example of how the buffer zones are sized for a part chosen for strategic replenishment.

Green Zone:

The average daily usage (ADU) for this part is 10 and the lead time is 7 days. This part has been placed in a medium lead-time category by the buyer. As such, the planner is using 50 percent of ADU × lead time. This will yield a calculated green zone of 35 units. While this part does have a minimum order quantity of 5 units, the calculated green zone is higher. Thus, the calculated green zone will be used. The reason for taking the higher is that the green zone essentially determines order frequency.

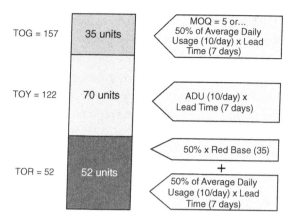

Figure A.2 Strategic Replenishment Buffer Sizing Example for Part XYZ

Setting the green zone to the MOQ would mean that on average this part would be ordered twice per day (remember the ADU is 10). This situation may prove to be too transitionally intensive, costly, and even capacity-erosive for an environment especially if this same scenario is spread over hundreds of parts. In other words, there can be a point of diminishing returns even to frequency of ordering, a point where flow can actually be impeded due to certain restrictions and limitations inherent in the system.

Yellow Zone:

The yellow zone for Part XYZ is calculated by multiplying the full ADU by the lead time (10/day × 7days). This yields a yellow zone of 70.

Red Zone:

Red zone base, in this case is going to mirror the green zone. The part is in the medium lead-time category and the planner has once again set the ADU percentage multiplier to 50 percent. This yields a red base of 35 units. Next the red safety is calculated as a percentage of the red base according to the variability category the part falls within. In this case the part is in the medium-variability category and a percentage of 50 percent is chosen. Red safety for Part XYZ is red base (35) × 50% = 17.5. In this case it has been rounded down to 17. The total red zone is 52 (red base [35] + red safety [17]). This number is also known as top of red or TOR.

The top of the buffer (called top of green or TOG) is the summation of these calculated zones. For Part XYZ top of green = 157. The top of yellow or TOY is 122 and equal to the TOR (52) + the yellow zone (70).

It is important to note that the top of green *does not* represent a targeted on-hand quantity. It simply represents the upper range of targeted available stock quantity. The targeted average on-hand quantity can be derived by adding half of the green zone plus the total red zone. This will be explored later in this Appendix.

Planned Adjustments

These buffer calculations are simply the starting point. The equations should yield a result that a company and its planning and buying personnel initially feel comfortable with. Now it is a question of making the buffers dynamic, giving them the ability to adapt to changes that are both experienced and planned.

These buffers are adjusted or "flexed" up or down based on changes or modifications to the ADU calculation. Figure A.3 represents different scenarios of adjustment. In all charts the dotted lines moving left to right correspond to the right y-axis and represent the ADU. The color-coded bars represent the quantity of the total buffer and its respective zones according to the left y-axis labeled "Quantity". The upper-left scenario with the title "Changes to calculated ADU over time" represents the gradual change to the average rate of use over 12 periods of time using a rolling average. The upper-right scenario represents a seasonal situation in which relying on the rolling past ADU may be insufficient to cover the seasonal upsurge. In this case the ADU is actually manipulated in advance by applying a percentage factor to it. Period 6 is set to an ADU of 180 percent while period 1 may be set to an ADU of 80 percent. This ADU manipulation is called a planned adjustment factor (PAF) and can pattern historical seasonal behavior, expected seasonal behavior, or some combination of both.

The lower-left chart in Fig. A.3 is a planned adjustment made for a part that is being taken out of service. By period 10 in the chart, the part will no longer be active. A glide path is created stepping the ADU down (through a decreasing ADU percentage equation) over the time range. The objective is to prevent large reorder quantities and the obsolescence risk associated with them so close to the end-of-service date.

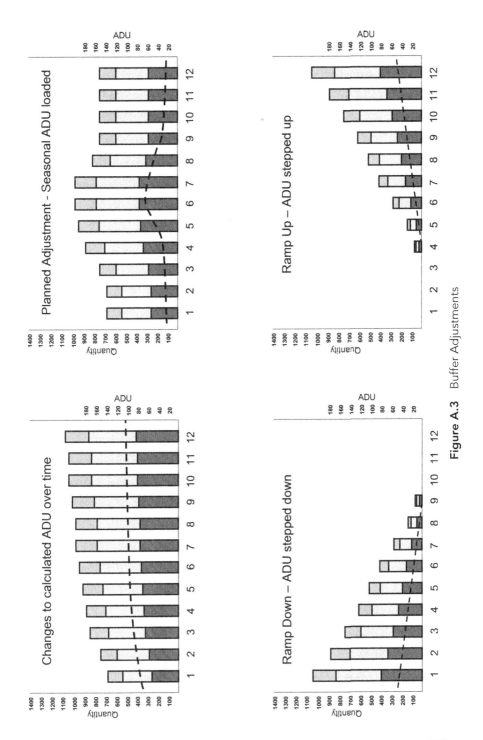

Figure A.3 Buffer Adjustments

293

The lower-right chart in Fig. A.3 represents a new part introduction scenario. The problems with new parts are that there is no usage history recorded and sometimes expected sales can be overstated. To address the first issue, we can use a forecasted ADU to start on Day 1 and then blend it with actual usage through a rolling calculation until that rolling calculation is comprised of purely historical usage. To address the second issue, a ramp up plan can be established that applies a percentage of that ADU over time stepping up the buffer levels. This is often done when a company chooses not to commit the capacity, cash, and/or space until the market is proven. The ramp up path provides a risk-averse way to test but still service the market.

Supply Order Generation

These buffers are replenished through the consideration of on-hand, on-order, and actual demand orders. Supply orders are generated based on what zone the available stock is in. The available stock equation for replenished finished items = On-hand + On-order (open supply) – Sales Order Demand (due in the past, due to today, and qualified spikes). The DDMRP available stock equation for intermediates and purchased items = On-hand + On-order (open supply) – Work Order Demand Allocations (due in the past, due to today, and qualified spikes).

Figure A.4 describes how order spikes are qualified. In order to qualify order spikes an order spike horizon is established. An order

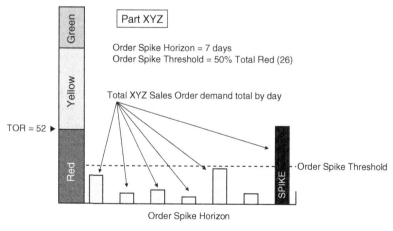

Figure A.4   Order Spike Qualification

spike horizon is at least one lead time in the future. In Fig. A.4 Part XYZ's order spike horizon is 7 days (corresponding directly to the part's lead time). Next an order spike threshold is established. In this example, it is set at 50 percent of the total red zone, 26 units. The order spike horizon looks into the future at the summation of a part's *sales orders* by day (for intermediate or purchased items it would be work orders). This means several sales orders due on the same day can accumulate to make a spike. In the above example, an accumulation of 26 units or greater, a little over two and a half days of average demand, on any single day will be declared a spike and compensated for by its inclusion in the available stock equation.

Order spike qualification and compensation gives the buffer better dampening capability without additional inventory investment.

Figure A.5 demonstrates the supply order generation process from strategically replenished buffers for a manufacturing plant with both machining and assembly capability. Decoupling points have been placed between the assembly operation and the customer by stocking finished items. The bucket icons to the right are finished stock positions (parts FPB, FPA, and FPE). Additionally, decoupling points have been placed between machining and assembly in the form of stocked intermediate items (parts ICB, SAF, and ICS).

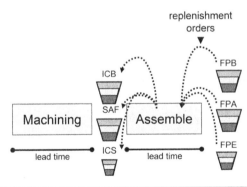

Part	Open Supply	On-hand	Demand	Available Stock	Recommended Supply Qty	Action
FPB	5453	4012	3200	6265 (49%)	6500	Create Order
FPE	3358	4054	540	6872 (66%)	3128	Create Order
FPA	530	3721	213	4038 (67%)	2162	Create Order

Figure A.5 Supply Order Generation Example

As described previously, order recommendation is based on the available stock equation. If the available stock equation yields a position below the top of yellow, then a supply order is generated for a quantity up to the top of the green zone. Part FPB has an available stock position of 6265. That number was derived by adding open supply of 5453 to on-hand of 4012 and subtracting qualified sales order demand of 3200. The available stock position is color coded with yellow, meaning it is in the yellow zone (below top of yellow but above top of red). The percentage indicator in the available stock column relates the current available stock position to the total buffer (top of green quantity). Available stock is 6265 or 49% of the top of green number and a supply order of 6500 units is recommended to restore the available stock position to the top of green.

Orders for resupply of finished items (labeled "replenishment orders") are represented by the dotted lines connecting back into the assembly operation. These replenishment orders will be converted to work orders and scheduled into assembly. The work orders scheduled in assembly represent demand allocations against the intermediate stock buffer but unless they are due for release today or represent a spike on a future day within the spike horizon, they do not impact their respective available stock equations.

As mentioned previously, the average on-hand inventory position is considerably less than the top of the green. The top of the green zone is simply an order-up-to level. Figure A.6 illustrates the buffer and relevant properties for a stocked part. We will use this example to demonstrate how to arrive at determining the average or target on-hand level. The lead time of this part is 18 days with an average daily usage of 10. The green zone is set to 30 percent of usage over lead time yielding a green zone of 60 units. The yellow zone (always 100 percent ADU over lead time) will be 180 units. Let's say that the total red zone is set at 100.

If the green zone is 60 units and the ADU is 10 units per day then in a perfectly average world, we will reorder 60 units every 6 days. If we reorder the part every 6 days, we will typically have 3 open supply orders at any one time when the available stock equation is green. We would have no more than three (assuming average demand) because the available stock would go over top of green or on-hand would have to be significantly depleted. Figure A.7 depicts the three open supply orders.

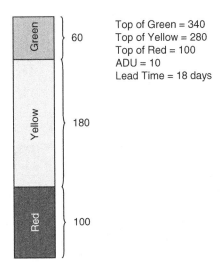

Top of Green = 340
Top of Yellow = 280
Top of Red = 100
ADU = 10
Lead Time = 18 days

Figure A.6 A Buffer Example

After completing the daily plan and accepting orders, a planner should have 3 open supply orders for at least 60 units each. With three open orders there would be at least 180 units of open supply when the available stock is at the top of green and on-hand should be no more than 160 pieces. We have 6 days of demand left until we receive an open supply order and launch a new supply order. ADU over 6 days would drain the on-hand quantity down to 100 before the receipt of an open supply order of 60 brings our on-hand level up to 160 again. In our perfectly average world our on-hand inventory would range between 100 and 160 with an average of 130. Total red (100) + half of green (30) = 130.

This is illustrated in Fig. A.8. The dark boxes moving in descending fashion from left to right depict daily usage (10 per day). Starting at 160 units, each day's demand drives the on-hand position down until a supply order is received. The receipt of the supply is indicated by the tall skinny rectangle.

DAY	1	2	3	4	5	6	7	8	9	10	11	12	13	14	15	16	17	18
Supply Order	60						60						60					

Figure A.7 Open Supply Orders Over Lead Time

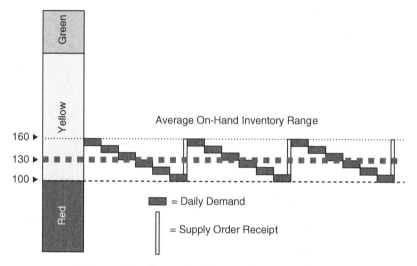

Figure A.8   Illustrating Average Inventory Range

As previously demonstrated in the equations above, replenishment buffers use only actual customer demand or consumption (typically represented in the form of sales orders). Intermediate component work order demand allocations are assumed to be driven by sales orders. Using only sales order demand means replenished positions are not susceptible to the forecast error that grossly inflates statistical safety of stock positions. Forecasted orders are simply not part of the supply generation equation.

So far we have only defined how and why replenishment supply order generation provides better planning signals. Strategically replenished buffers also provide a huge leap forward for execution visibility—the management of open supply orders. In conventional formal planning, priority is determined by due date. That is all there is to see. But situations change, especially in the New Normal. Delays and breakdowns happen. Consumption spikes occur. Quality and credit holds are placed. All of these can have huge implications for the real priorities of existing supply orders. Today, we are relatively blind to the impact of these changes on priority. We do not have the relevant data, or data is hidden in huge queues of MRP action flags.

The strategically replenished buffers of DDMRP provide powerful execution tools. These buffers bring a whole new aspect of execution to light—priority by buffer status.

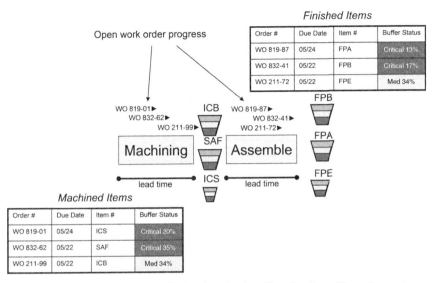

Finished Items

Order #	Due Date	Item #	Buffer Status
WO 819-87	05/24	FPA	Critical 13%
WO 832-41	05/22	FPB	Critical 17%
WO 211-72	05/22	FPE	Med 34%

Open work order progress

WO 819-01 ▶
WO 832-62 ▶
WO 211-99 ▶

ICB

WO 819-87 ▶
WO 832-41 ▶
WO 211-72 ▶

FPB

SAF

Machining

Assemble

FPA

lead time

ICS

lead time

FPE

Machined Items

Order #	Due Date	Item #	Buffer Status
WO 819-01	05/24	ICS	Critical 20%
WO 832-62	05/22	SAF	Critical 35%
WO 211-99	05/22	ICB	Med 34%

Figure A.9 Using Strategically Replenished Buffers for Shop Floor Execution

Figure A.9 represents the new kind of visibility that comes with these types of buffers. From left to right, the two tables display the order number, the due date of the order to the buffer, the item number, and the on-hand inventory status of the part's buffer. There is work-in-process in the form of work orders in both machining and assembly. Assembly has three work orders. The relative position is meant to convey how far each work order has progressed through the process. The closer to the buffer on the right, the farther along the work order is. For example, Work Order 832-41 is nearest to completion.

The table, frequently referred to as a *current inventory alert board*, indicates Work Order 819-87 has the highest priority from an on-hand perspective but the latest due date. The current inventory alert visibility allows the floor to shift activity to protect stock buffers with the highest priority need.

These buffers have been strategically chosen as decoupling points and are often the heart of a demand driven system. They should not be thought of as simply "stock". They also are strategically stored capacity and time. To that extent they are all three types of buffers (stock, capacity and time) rolled into one, and a very important systemic control and variability dampening mechanism.

INDEX

CPSIA information can be obtained
at www.ICGtesting.com
Printed in the USA
LVOW13*1048310718

584925LV00002BA/34/P